WALKING EUROPE'S LAST WILDERNESS

WALKING EUROPE'S LAST WILDERNESS

A Journey through the Carpathian Mountains

NICK THORPE

YALE UNIVERSITY PRESS
NEW HAVEN AND LONDON

Published with assistance from the Charles S. Brooks Fund.

For information about this and other Yale University Press publications, please contact:
U.S. Office: sales.press@yale.edu yalebooks.com
Europe Office: sales@yaleup.co.uk yalebooks.co.uk

Set in Adobe Garamond Pro by IDSUK (DataConnection) Ltd
Printed in Great Britain by TJ Books, Padstow, Cornwall

Library of Congress Control Number: 2024947846

ISBN 978-0-300-25354-2

A catalogue record for this book is available from the British Library.

10 9 8 7 6 5 4 3 2 1

This book is for Janet, my mother

The Quiet Miracle of Motherhood

Nothing new from the workshop of God:
Love makes a man vulnerable.

Out of this spirit is made such a holiday,
such a great holiday that the pain is worth it.

Love – it's not a child of the moment.
Love is long, life is short.

Do you not understand that hidden power?
Then ask your mother.

Mother is a well; you are a bucket.
A well full of living water.

That you could draw from
for a thousand years
and not exhaust.

So kneel before her

And only after may you leave.

Milan Rúfus (trans. Allan Stevo)

CONTENTS

CONTENTS

ILLUSTRATIONS

PLATES

13. Vasyl and his cheeses, 2021. Author's photograph © Nick Thorpe.
14. Vasyl Kischuk, 2021. Author's photograph © Nick Thorpe.
15. *Daybreak in the Retezat Mountains* by Andrea Weichinger, 2023. © Andrea Weichinger.

MAPS

ACKNOWLEDGEMENTS

This book is based on travels over six years, from 2018 to 2024, alone or with friends and family. Sometimes the people concerned are named; sometimes they are not, in the interests of simplicity. Thank you all for your companionship and help.

In Slovakia, I would like to thank in particular Palo Littera, who introduced me to his amazing network of friends and acquaintances, working to protect or restore the landscape. Many of their names appear in this narrative. In the High Tatras, Martin Mikoláš was a constant source of enthusiasm and ideas.

In the Czech Republic, I am grateful to Michal Medek, author and traveller, for sharing his vast knowledge and encouraging me to write a unifying narrative of the much-fragmented Carpathians.

In Transcarpathian Ukraine, I would like to thank especially Alysa Smyrna, Oleksandr and Gabriela Nikitchuk, Taras Bigun, Serhii Prokop, Márta Palojtay, Oreste Del Sol, Peter Popovics, Valentyn Voloshyn, Vasyl Pokynchereda, Viktor Stenich, Rostislav Martinyuk and Vasyl Losiuk.

In Romania, Mircea Barbu, Feri Ségercz, Alison Mutler, Ion Holban, Cristi Papp, Tibor Kálnoky, Ciprian Gălușca, Gabi Paun, Ioana Raluca Voicu-Arnăuțoiu and Chris Mititelu.

Among the many people who inspired this account, I would like to mention the life work of ecologist, poet and author Gary Snyder, whom I was fortunate to meet once, many moons ago, at a book-signing in Bristol. The British author Robert Macfarlane has also been an important inspiration, in particular his first book, *The Mountains of the Mind*. My friend Julius Strauss, bear guide and reporter, taught me how to approach (or not approach) bears, and was a great source of inspiration and advice.

My friends Roger Norman, Steve Batty and Sarah Smith all moved on up the mountain path while I was writing. I miss them hugely.

My wife Andrea travelled with me often, and put up with my long disappearances in the hills. Featuring pigments mixed with egg tempera on khadi paper, her paintings of the Carpathians capture the purples and blues and greens of mountains and forests beautifully, where words stumble.

As well as sharing the wilds of Transylvania with me, our sons Jack, Caspar, Daniel, Máté and Sam are travelling the world themselves now – from the Scottish Highlands to the high Andes, Turkey, the Caucasus, the Himalayas, Central Asia, the coastlands of Spain and Portugal. As I write this, Sam is sailing from New Zealand to Japan on a 16-metre yacht, travelling at 6.4 knots in a strengthening easterly wind off New Caledonia, heading towards the Solomon Islands.

My sister Mish braved the Retezat, the Făgăraş and the Ţarcu mountains and came with us all the way to the Iron Gates, while my brother Dom listened patiently to our unlikely tales.

This book is dedicated to our mother, Janet, who passed away in November 2022, aged ninety-six. All mountains remind me of our family journeys on Ide Hill and Toys Hill, near Sevenoaks in Kent, and on the Long Mynd in Shropshire, where I learnt to walk, and where my father, Peter, showed me the way the wood on the stiles had been rubbed smooth by so many hands.

At the BBC I would like to thank numerous colleagues, but in particular Pete Karlsen, Paul Kirby, Bethany Bell, Orsi Szoboszlay, Polly

Hope, Tim Mansel, Richard Colebourn and Penny Murphy. At Yale University Press, my thanks in particular to Julian Loose for his faith and encouragement, and to Rachael Lonsdale and Frazer Martin for guiding the text all the way to the printed page.

Michael Ignatieff and James de Candole read early versions of the text and offered valuable suggestions. Rachel Christoffel and Gyuri Kékessy, the Banjos, Gábor Máté and Gergö Somogyvári were the best friends and allies a travelling man might ever need.

Last, but by no means least, my gratitude to fellow Hungarian Clive Liddiard, for restoring order and simplicity to the rocky slopes of my mountainous grammar.

NEW SONGS FROM THE WEST

Hath not old custom made this life more sweet
Than that of painted pomp? Are not these woods
More free from peril than the envious court?

> Duke Senior, in William Shakespeare, *As You Like It*

It's true they weren't mine according to the land register. But they were mine according to God's will, because I was born there and that's where I became a man. Give me back my mountains!

> Albert Wass (1908–98), Hungarian writer

Imagine a grand piano, on which all the visible keys and the strings buried beneath the soundboard are streams and rivers, and the body of the instrument is the mountains. Then imagine the piano is shaped like a horseshoe, with the keys arranged, not just black and white, but all the shades of greens and browns, yellows and reds, blues and greys, spread from end to end all the way around the instrument, on the inside as well as the outside rim. Then remember that a mountain absorbs but does not emit light. The piano rises dark and promising at the geographical centre of Europe, with the lights of towns and villages and cities flickering all around it.

In this book, I would like to invite the people who live in and among the Carpathians to play it for you. I am just the man in the car park, from whom you ask the way.

⊙ ⊙ ⊙

The Carpathians cut through the former Austro-Hungarian empire like a boomerang. On the fault line between East and West, the mountains anchor these lands at the geographical centre of Europe. On satellite images of the earth at night, the Carpathians are easily distinguished as a dark ring, between the bright lights of Vienna, Bratislava, Budapest, Brno, Kraków, Lviv and Bucharest.

Those who believe in walls and fences between peoples for ideological or tribal reasons have a field day here. In the Cold War, people who tried to cross the Morava to Austria were shot by Czechoslovak border guards. There's a monument to the victims on the shore at Devín: a white gate riddled with bullet holes. The iron bars across the gateway have been partially broken, to symbolise the victory of freedom:

> You, who are free and unrestrained, shall not forget that freedom of thinking, action and dreaming is a value worth living for and sacrifice.

Between 1945 and 1989, 400 people lost their lives trying to cross the Iron Curtain from Czechoslovakia to Austria or Germany. Nowadays, there are not supposed to be customs controls between Austria, Slovakia and Hungary, as they all belong to the Schengen visa-free zone. But Austria and Slovakia regularly reinstate border checks, because they have noticed how many migrants and asylum seekers mysteriously enter their countries through Hungary, whose government claims to stop them all at the Serbian or Romanian border.

My journey begins at Devín, half an hour's walk east of the Slovak capital Bratislava, where a cliff topped by a ruined castle slopes dramatically down to the River Danube, like the jagged wings of a cockerel. To the west of the cliff, the Morava flows into the Danube, 350km from its source on the Czech–Polish border, skirting the rim of the mountains. Willows overhang the banks, and plastic bags and bottles are stranded among the roots. The dark stain of the Morava spreads out far into the blonder, sandier Danube.

At the top of the cliff are the ruins of a castle, blown up by Napoleon in 1809 and never repaired. Devín in Slovak, Dévény in Hungarian, Theben in German, a small place with three names. This is true of almost everywhere I refer to in this book. Rivers change names as they cross borders: the winding Tysa shifts to Tisza then Tisa. I try to solve this problem by using the name used by the majority of the population in each place, but majorities change with migrations and deportations. I apologise in advance to anyone who feels offended by my choice. On the maps provided, I have stuck to the regular atlas spelling.

Devín was the gateway between Hungary and the West. When the Magyar tribes arrived in the ninth century, they found the Moravians well established here. The Magyars overcame them, and launched raids deep into western Europe. They only withdrew to Devín after a catastrophic defeat by King Otto the Great of Bavaria at the Battle of Lechfeld in 955.

Devín has also served as a gateway of the imagination:

I came along Verecke's famous path,
old Magyar tunes still rage in my breast –
will it arouse your Lordships' righteous wrath
as I burst in at Dévény, with new songs from the West?

Written in 1906 by the Hungarian poet Endre Ady, the author targets the 'lordships' of both Hungary and Austria for what he sees as their stultifying grip on culture.

The roar of mountain streams, the rattle of woodpeckers, and the strange popping sound of the capercaillie, *Tetrao urogallus*, tempted the poets into these woods. I quote a lot of poets, because whatever their political views, at least they revered the trees.

The Carpathians are the watershed between the Baltic Sea in the north and the Black Sea in the southeast. The Vistula, Poland's greatest river, whose name comes from a Sanskrit root meaning 'to ooze, or flow slowly', rises on Barania Góra (Sheep Mountain) in the western Carpathians, to flow a thousand kilometres to the Baltic at Gdańsk.

I follow the mountains clockwise, 1,500km from the Danube at Devín, to the Danube at Orşova, on the Romanian border with Serbia. In *The Danube: A journey upriver from the Black Sea to the Black Forest*, I reached the westernmost well-springs, the furthest fingertips of the Danube, but realised that her muscles are to be found in the Carpathians. This is my latest attempt to find out 'where the waters come from'. It is not a guide-book for mountain climbers. I walk a lot, up hill and down dale, but climb few peaks – I was more interested in the people who call these mountains home than in those brave souls who travel here to pit themselves against the caprices of snow and avalanche. The Hungarian word *pálinka*, a tradi-tional fruit spirit, crops up often in the text, thanks to the warm hospitality of those I met on my journey. I have chosen not to translate it each time.

Occasionally, reference is made to previous trips through the moun-tains, for example in the winter of 1986 and the summer of 1993. The careful reader will notice that the seasons sometimes change abruptly, or occasionally do not follow each other in the usual order. This is a consequence of compressing several years of travel into a single narra-tive. During those years, a pandemic and a war made travel more diffi-cult, but not impossible. Mountains, after all, are a place of healing, for all the troubles of the body and the soul.

'They have no history,' an anonymous nineteenth-century British author wrote of the Carpathians, because he was too lazy to go there to find out. The Carpathians actually groan under the weight of their history. The old tribes of Europe – Celts, Romans, Germans, Franks, Slavs, Huns, Dacians, Avars and Magyars – set up camps in their foot-hills and strongholds on their vantage points. Many valued them for their salt. The medieval kings and queens of Europe treasured them for their copper, silver and gold.

Why have we heard so little about the Carpathians? Largely because they are so varied, each massif different to the next. On the map they look more like a single range. Up close they reminded me often of a pack of cards, constantly reshuffled, as I found my way between them. For Poles, they are a southern border; for Ukrainians, a western border-land; for Slovaks, the three stylised mountain ranges on their coat of arms. Rather than the Carpathians, Hungarians speak of the Carpathian *basin*, their homeland, and still mourn the loss of the ring of mountains that once protected them from the outside world. During the writing of this book, I became a Hungarian citizen and somehow inherited their pain. But I cross the spine of the mountains at every opportunity, like the roof of a house, to experience what it looks like from the other side. For Romanians, the mountains are central to their image of them-selves, as descendants of the forced marriage between the brave Dacians and their Roman conquerors. The mountains also hide less well-known identities: Ruthenians and Transylvanians, Liptos, Lemkos, Boykos and Hutsuls, Bukovinians and Szeklers.

'Which is your favourite mountain?' I asked Serhii Prokop, a moun-tain guide in Transcarpathia in Ukraine. 'Pip Ivan in the Marmaroshchyna range,' he replied, 'but my favourite view is from my village on the plains of the Tysa river, looking up at the distant peaks.'

Who are you? I asked Anna, aged ninety-five, a resident of a little village in the eastern Carpathians, beneath Ukraine's highest peak, Hoverla. 'I'm a Hutsul from the Bohdan valley,' she replied. Her daughter Ahava, aged seventy-eight, described herself as a Hutsul from the Breboja region. The Hutsuls are a tribe or ethnic group, so old they have their own origin myth. God and the Devil decided to make the world together. The Devil kept lying down to rest, and in his anger, God kicked him. And each time the Devil rolled over, a bit more land appeared beneath him.

The older a person is, I notice, the narrower the place they identify as home.

Wherever I travel, I ask people about their identity and where they call home – the particular landscape of mountain or valley. This often chimes with the idea of a nation – but not always.

There is a fierce academic debate about the ancient or modern origin of nations. Ernest Gellner suggested that the nation and nationalism were the products of industrialisation and the war waged by France after the 1789 revolution. Anthony Smith proposed the existence of a 'primordial nationalism'. 'Nations have navels,' his followers wrote, and did not 'spring fully developed from the new nation-state'.

Nations have long existed as 'imagined communities', wrote Benedict Anderson, as people in one place could imagine other people far away in the same kingdom whom they would never meet, but with whom they felt they had something in common. They respected the monarch and worshipped God in the same way. Those unseen folk beyond the hills were somehow their kin.

Anderson believed that 'imagined communities' could be traced back to the printing press. Books helped transform something that was already present as a dull feeling of belonging into a firm idea, which people were willing to die for. This was brought into sharp focus by the full-scale Russian invasion of Ukraine in February 2022. In Bosnia, in the early 1990s, I witnessed a country both simultaneously broken apart and forged in war. Similarly, the Ukrainian identity, distinct from Russia, was forged in the Holodomor, the terrible starvation imposed by Joseph Stalin in the early 1930s, and again by the Russian invasion since 2014.

The historian Tara Zahra introduced the interesting notion of 'national indifference'. 'What would you advise your son if he returned from Switzerland to Ukraine?' I asked Jolána, an ethnic Hungarian from Transcarpathian Ukraine, in the spring of 2023. 'I would tell him he should live for his country, not die for it,' she replied quietly. Nearby, I drank small glasses of *horilka*, a powerful spirit flavoured on this occasion with hawthorn root, with a farmer who identified as Ruthene, rather than as Ukrainian.

Down the road in Khust, a Hungarian woman in a shop said she was relieved that her own son was abroad. 'But thank God some boys remained to fight,' she added. 'We know what it's like when Moscow is in power, and we never want that again.' There are contradictions in all our stories.

In the course of writing this book, I corresponded with my former colleague at the *Observer* newspaper in London, the Scottish writer and journalist Neal Ascherson. In his book *Stone Voices* on Scottish history, he quotes the 1320 Declaration of Arbroath: 'While a hundred of us remain alive, we will never submit to the domination of the English' and never 'give up freedom which no good man abandons except with his life'.

'I used to think it was a forgery,' Neal wrote to me,

because it identified national freedom with individual freedom, and it seemed far too early for ideas like that. But now I see it differently. It's obvious really – I'm an unfree peasant or villein working my patch of earth for a feudal lord, in a familiar, beloved landscape, but it's a bleak life. On the other hand, when the English (or Turkish or German) soldiers invade this kingdom, it's my house burned, my cow and geese stolen and killed, my daughter raped. So I'll fight for Scotland's freedom simply because it's bound up with mine.

What both nationalists and globalists like to forget is that their creeds are recent intruders in the grand scheme of things. The nation-states of Europe were created in the nineteenth century by unravelling the complex web of religious, cultural and linguistic threads that characterised Europe, and which had both engendered and solved her previous wars and conflicts. Local dialects, vernaculars and identities were targeted for eradication in the service of the nation. The nationalists declared war on local differences before they went after foreigners.

The mountains are a place of refuge. Their folklore is rich with the exploits of bandits – like the Slovak Juraj Jánošík or the Ukrainian

Oleksa Dovbush. The western Carpathians in Slovakia gave refuge to the communist partisans who fought the Nazis, and the southern Carpathians in Romania sheltered the partisans who fought the communists. Ukrainian partisans fought the Poles and the Soviets here. Hungarian soldiers built the Árpád Line – three layers of resistance against the invading Red Army in August 1944 – but were forced to abandon their defences when Romania switched sides and the Russians drove their tanks round the back.

The mountains are home to plants that will heal your ailments (if you can find someone who still knows how to brew them up and the appropriate phase of the moon to do it by). The bark of the linden heals inflammation on the legs of sheep. The fruit of the service tree soothes the swollen guts of humans, cows and goats. In the guesthouse beside the Sucevița monastery in Bukovina we ate cep mushrooms, night after night. 'Where do you get so many from?' I asked the manager, innocently.

'An elderly lady arrived with baskets full of them last autumn,' he said. 'When there are so many mushrooms, she warned me, it means war is coming.' Within four months, the Russians had advanced deep into Ukraine.

On my shelves in Budapest, I have a large volume called *Laments*, in Hungarian and English, the verses and musical scores of all the laments for the dead collected by the composer Zoltán Kodály in tiny villages on the eve of the First World War. At the back is a pull-out map. The places where the laments were found and recorded are marked in purple, the rivers in grey, like the tracks of tears. There are no roads.

The towns and villages and cities of Europe glow at night like clock faces in a shop window, or flicker like screens, a herd of mobile phones grazing in a shopping mall. My eyes hurt from staring too long at the screen and the colours blur. I close my eyes. As the white noise, the chatter of sleepless machines subsides, I remember nights in the mountains when the clouds cleared and the stars filled a universe framed by the jagged chins and ragged ears of the hills. And the scent of wildflowers, and the night cries of birds.

The land does not belong to us, we belong to the land, I believe. The Carpathians are changing, and this book is both a chronicle of those changes and an appeal to preserve what can still be saved. The upland meadows of Slovakia and Poland have been lost because there are no more sheep to keep back the shrubs and trees of the encroaching forest.

In Ukraine and Romania, the meadows are disappearing fast. There are fewer and fewer young men and women willing to endure the hardship of life on the land. One Romanian in four now works in western Europe.

Animals make the landscape, as the governments of Austria and Switzerland know well, as they subsidise rural communities to graze their animals where visitors from the cities can admire them. Romania advertises itself as the 'green heart of Europe' – one of the last places where you can find a hundred species of wildflowers on a short walk. But Romania's small-scale farmers have been abandoned by the state. If they were supported, more would stay.

More must be done to save the last virgin and old-growth forests of Slovakia, Romania and Ukraine. Beech trees over 200 years old are being cut down. Romanian foresters are engaged in a race against time to cut as much as they can before their own laws are tightened. In the High Tatras in Slovakia, it took environmentalists twenty years to save the Tichá and Kôprová valleys from the chainsaw. Today they are among the most awe-inspiring places on earth.

There is a Carpathian Convention, a Carpathian Foundation, a Carpathian long-distance path and many other worthy initiatives, but little awareness of the whole magnificent horseshoe. This book aims to put the Carpathians on the map.

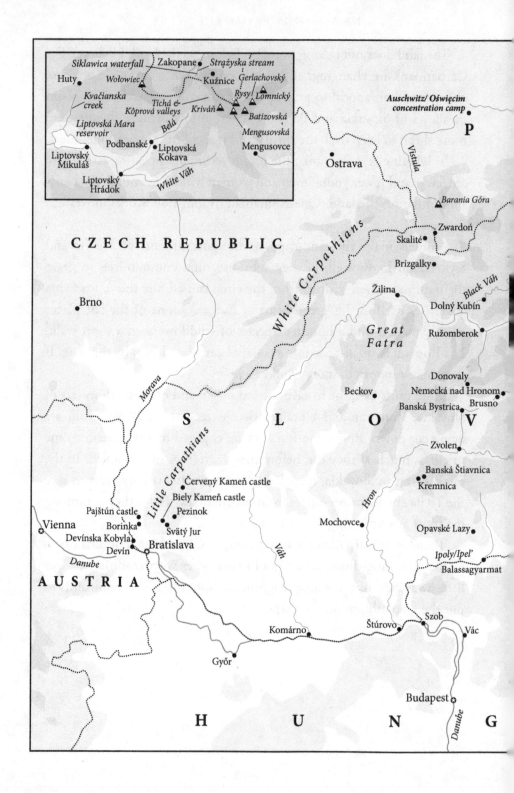

P O L A N D

Vistula

Kraków

Dunajec

Jasło

Black
Dunajec   Nowy   White
błonka   Targ   Dunajec

Wisłoczek

Dukla pass

San

Podhale
district

Červený Kláštor

Svidník

Cisna

Lutowiska

Przysłup   Smerek

Poprad

Strážky

Pusté Pole

Smolnik

High Tatras

Okraglik   Wetlyna

inset above left

Poprad

Prešov

ow Tatras

Kráľova hoľa

Iron
nie   Ďurková
cabin

Telgárt

Hnilec

Michalovce

Uzhhorod

A   K   I   A

UKRAINE

Hidasnémeti

Košice

Bodrog

Miskolc

Hornáď/Hernád

Tokaj

Tisza

Debrecen

N

A   R   Y

0   40 miles
0   40 km

CHAPTER 1

ADONIS IN THE GRASSLANDS

Each nation has a centre of happiness within itself, just as every sphere has a centre of gravity.

Johann Gottfried Herder (1744–1803), German philosopher

New research suggests that the standard practice of using managed honey bees to pollinate commercial apple orchards may in fact be unnecessary – wild bees may be just as effective, and produce better quality fruit.

European Commission, December 2022

The idea of the nation reminds me of domesticated bees: useful as far as they go, but what we really need is wild bees. Humans, like bees, are local creatures, drawn back to a certain valley, street or doorway. We need a sense of home, of a geographical place in the world to which we can return.

Palo Littera walks ahead with his dog Lenka up the steep path at Devínska Kobyla, the village beside Devín. It's a sunny afternoon in late March. A riot of bright yellow flowers, like giant buttercups, spatters the hillside ahead of us. *Adonis vernalis*, yellow pheasant's eye – an upright flower with up to twenty petals, rising from a groin of shy, curly leaves – covers the whole slope. It only blooms for two weeks a year. 'Adonis was a mythical character, the prettiest boy in the world,' explains Palo, an ecologist with a straggly ginger beard, hair tied back in a pony-tail and a thin, weather-beaten face lit up by a broad grin. 'He was very proud of his beauty, but as in ancient times there were no mirrors, he spent a lot of time by the shore of a lake enjoying his reflection. One day he slipped and drowned. And this beautiful flower was named after him.'

We sit for a while in silence on the hillside, looking down on the flowers. The spring sun is already warm on our faces, but there's a chill breeze. From a few rare specimens, the flowers have multiplied here in the last ten years into an ocean of adonis, and their fame has spread. Families with children in pushchairs or baby-carriers, couples holding hands and lone hikers with dogs form a steady procession along the path.

> The *Adonis vernalis* are actually poisonous. That's why the goats, who are very intelligent animals, don't eat them. But they eat all around them, which keeps the grass down. These flowers need sunlight, so they flourish.

The biggest animals which grazed these ancient grasslands were the straight-tusked elephants, *Palaeoloxodon antiquus*, 4 metres high at the shoulder, weighing up to 11 tonnes, driven to extinction some 28,000 years ago. There were also large herds of wild horses, aurochs and water buffalo.

> Imagine the kind of savanna landscape you see nowadays in Africa, but even richer in species. The elephants were the architects which formed the landscape and maintained the grasslands.

Hunting and climate change eventually destroyed the elephants, and domesticated animals like sheep and goats graze the grasslands instead.

A thousand varieties of wildflowers, herbs and grasses flourish here, thanks to the grazers.

BROZ, the environmental group Palo works for, is cutting down the trees that the communist authorities planted here sixty years ago. 'They weren't malevolent, or stupid. They just wanted to make the hillside "productive" in their own terms, to get some timber out of it.'

The communists' tree of choice was the flowering ash, *Fraxinus ornus*, a native of southern Europe. What they didn't realise was that, by planting trees here, they would destroy the ancient grasslands. Further along the path we reach a patch of woodland, not yet cleared by Palo's team. Last autumn's leaves are still thick underfoot. There are no flowers or grasses.

On the open hillside, we kneel beside the path. Purple mountain pasque flowers, *Pulsatilla montana*, grow next to a clump of *adonis*. They like stony meadows, dry and lime-rich soil.

The quieter we are and the closer we look, the more species emerge from the hillside. A small black male beetle clings stoically to a much larger female, loaded with eggs.

Some of these insects mate for days or weeks on end. The female beetles lay their eggs on flowers like this one. When the larva comes out from the egg, it waits on the flower. And these flowers are pollinated by solitary bees. When the bee comes, the larva stick to it. And it lets the bee carry it to its nest. The larva develops inside the bees' nest and steals their nectar. It's basically a parasite. Then the mature beetles emerge.

A long barge negotiates the broad bend of the Danube below Devín castle, where the powerful currents of the Morava flow into the Danube. The sound of its engines comes and goes on the breeze. It's the *Mirella*, flying the Slovak flag, bound for Ybbs in Austria with a cargo of grain.

Wild bees swarm along the path, some barely half the size of my fingernail, in and out of small bee holes in the soil. 'Many people think that to save nature you should breed domestic bees. But these wild ones are much better pollinators. So it makes more sense to preserve natural grasslands than to keep bees.'

Palo shows me a tiny wasp on the back of a small white snail shell. 'In my language these are called the golden wasps, because they have such beautiful colours. Actually they are poisonous, and have a very interesting way of life.' As he speaks, the wasp crawls over the snail shell. A perfect vantage point from which to watch for her prey.

The golden wasps dig their nests in the ground like the bees, and make a chamber at the end. But they don't use nectar or pollen from flowers to feed their larvae. Instead they collect spiders. First they have to fight with the spiders, because the spiders are also predators. And when they win the fight, they hit the spider's backbone with their sting, which paralyses but doesn't kill it. It's still alive, but cannot move. The wasps cut the legs off the spiders, then take the living body to the end of the chamber. Then they lay their eggs on it. And when the larvae hatch, they slowly eat the spider alive. And when they finish their development, the adult wasps come out, and they're beautiful, like this one.

Palo unpadlocks a gate and we climb past a wooden stall built for the goats to find shade from the heat in summer, and to cluster for warmth in winter. They're kept down in the valley for now, because they would eat the pasque flowers; but they will be allowed back up onto the slopes later in the spring, when other, more common blue flowers take their place.

We taste a tiny, jagged green leaf, *Teucrium*, used in homeopathic remedies to treat intestinal worms. The goats know the different properties of the herbs and choose the ones they need when they get sick. There are roe deer droppings along the edge of the woods.

To our left, the castle of Devín stands at the top of the cliffs. Beyond that, on the far side of the Danube, are the Hundsheimer hills, the brief

Austrian part of the Carpathians. Beyond them, a plain, 70km wide, is dotted with turbines, catching the wind that blows down from the Alps through the Danube corridor, eastwards into Hungary.

Beyond the plain we see the Alps clearly, the sunlight glinting off their snowy foreheads. The tallest is Hochschwab at 2,277 metres, then a ridge of other peaks – all over 2,000 metres. 'This is all actually just one mountain range,' explains Palo. 'It starts in the Pyrenees in the Iberian peninsula, then continues through the Alps to the Carpathians, then the Balkan mountains, through to the Turkish mountains, the Caucasus, the Himalayas, and beyond.' This is known by geologists as the Alpide belt.

◎ ◎ ◎

Down in the village, it's warm enough to sit at an outside table at U srnčíka – 'At the Roe Deer'. Specialities on the menu include spinach pierogi stuffed with sheep's cheese, sour cream and fried onions, and farmers' dumplings filled with smoked meat and sour cabbage, garnished with fried onion rings. Slovakia is dumpling and onion country.

The first vines were probably planted along the southeastern slopes of the mountains by the Romans. In the Middle Ages, wines from the Little Carpathians were especially prized in the royal houses of Europe. But in the 1860s and 1870s an outbreak of the dreaded phylloxera disease destroyed three quarters of Europe's vineyards, including these. The *Phylloxera vastatrix* is an insect which bores into the roots of vines, causing them to wilt and die.

Sándor Haraszthy, a Hungarian winegrower who emigrated to the United States in 1840, may have contributed to the spread of the disease. On two trips back to Europe from California in 1849 and 1861 Haraszthy travelled widely, collecting cuttings to take back to the New World. He also brought with him American cuttings, which he hoped might help European winegrowers fight powdery mildew, which attacks the leaves of grapes. The first cases of phylloxera in France were identified soon after Haraszthy's visit in 1861. Within five years of his return to California, his own vineyards were ruined by the disease. Did he carry it with him on his boots or in his cuttings?

After the insect wiped out the vineyards in Devín, local people planted blackcurrant and redcurrant bushes in their place. By the eve of the First World War, the new bushes were flourishing and the berries were being exported in large quantities upriver to the markets of Vienna, and downriver to those of Budapest.

Then came war. The Austro-Hungarian empire fought on the side of Germany, and was broken apart in defeat. Upper Hungary, including Devín, became part of the new country of Czechoslovakia. The old markets were lost and something else had to be done with all those currants.

In 1922, local entrepreneur Alois Sonntag struck on the idea of blackcurrant wine. The labyrinth of wine cellars beneath the houses had been gathering cobwebs since the end of wine-making, forty years earlier. Now they were dusted down and swept out, and the old barrels replaced. Sitting on the benches of the Theben blackcurrant winery (the German name for Devín), the blackcurrant cuvée brewed by Sonntag's descendants is surprisingly dry, like a good merlot. In winter, they serve it hot, as *ríbezlák*.

◎ ◎ ◎

On Sunday, 24 April 1836 (St George's day in this part of the world), Ľudovít Štúr was just twenty years old, a starry-eyed poet with an uncertain beard, who met up at the foot of the castle hill at Devín with sixteen friends. So concerned were they that the police might notice them and prevent their 'patriotic excursion' that they had travelled separately, in small groups, to the rendezvous point. From there they set off in high spirits up the hill to the ruined castle. At that time, what is now Slovakia was known as Upper Hungary – a region of rivers and mountains, where Latin, German and Hungarian were the languages of the rulers, and where the main city lacked a Slovak name: Pressburg in German, Pozsony in Hungarian only became Bratislava in 1919.

The valleys and cities of the plain might be in the hands of the Austrian or Hungarian occupiers, as Štúr and his comrades saw it, but the youth could reclaim the heights, the holy places of the ancient Slav tribes and kingdoms – not by force of arms, but with poetry and song. At the top, they took off their hats and sang Štúr's new poem, to the tune of a well-known folk ballad 'Nitra, milá Nitra' ('Nitra, dear Nitra'):

> Devín, dear Devín, orphaned in the castle,
> tell us, when did your walls stand?
> – My walls stood in the time of Rastislav:
> the glory of your fathers, man of my walls . . .

After a short speech in which the young Ľudovít extolled the glories of the Great Moravian empire, which flourished along the Morava river before the Magyars arrived in the ninth century, he suggested that each of those present add a traditional Slavic name to his (or her) own. He chose Velislav. Then they all pledged an oath of allegiance to the Slavic nation, sang a popular Slovak anthem, 'Hey Slovaks', and walked back down the hill to the pub.

Štúr and his friends were deeply influenced by Johann Herder, the eighteenth-century German philosopher. Herder glorified 'das Volk' – the 'folk' – and proposed that all peoples should set about collecting their own folk songs, legends and dances, in order to establish their own identities. If they failed to do so, he warned, their precious languages, cultures and traditions might be lost, assimilated by larger, more aggressive ethnic groups nearby. For the Slavs of central Europe, that meant the Germans and Hungarians. For the Hungarians, it meant the Slavs and the Latins – the Romanians.

Herder understood the nation as a linguistic and cultural expression, rather than a political one, and opposed the idea of the nation-state. For him, the nation was something broad and ultimately benevolent. If peoples were only free to express themselves fully, he believed, they would not make war on one another. But after his death in 1803, his writings were taken up by the poets and political leaders of the emerging nationalities of the empire, to forge the idea of nation into a sword, as well as a shield.

In 1896, the Hungarian authorities erected seven monuments at strategic points on the compass, to mark the thousandth anniversary of the Hungarians' arrival in the Carpathian basin. Their aim was to strengthen loyalty to the Hungarian nation – among both Hungarians and national minorities. The statue of the warrior prince Arpád, looking west, with shield and sword resting at his side, towered on a plinth above the castle at Devín. Other monuments were unveiled on prominent hills at Nitra, Mukachevo, Brasov, Zemun, Pannonhalma and Pusztaszer, all meant to last a thousand years. After Hungary was dismembered by the 1920 Trianon treaty following the First World War, the five outside the boundaries of the rump state of Hungary were blown up or torn down.

◎ ◎ ◎

Mikuláš Huba suggests we meet in a basement café near the historic centre of Bratislava. He's a big, gentle man with a deep voice, touching seventy. A geographer and ecologist, he played an important part in the semi-underground environmental movement of the 1970s and 1980s in Czechoslovakia, which helped topple communist rule.

We are the only guests in the cavernous, dimly lit basement. The round tables and comfortable chairs summon a slightly subversive atmosphere, reminiscent of the 1980s. Only the thick cigarette smoke is missing.

His parents hiked the High Tatras, and his childhood holidays were spent in a 'very small, but very romantic' cottage in eastern Slovakia, near the Hornád river.

Whereas in the Czech part of Czechoslovakia, the communist authorities made life very difficult for dissidents, Slovakia was more relaxed. This was partly also because the 'resistance' was less demonstrably political, and more environmental. Those in power underestimated the importance of environmental issues. The communist ethos was to industrialise and urbanise Bratislava, and to build tower blocks to house the millions of peasants who were supposed to leave their villages and move to the new factories of the cities.

Thousands of volunteers spent their weekends and summer holidays enthusiastically repairing old houses, railway stations, monasteries and castles abandoned by the state because they didn't fit into the 'socialist dream' of human progress. Old buildings linked to the German- or Hungarian-speaking past were supposed to crumble quietly from view. To halt or reverse that process was deeply subversive, and by the time the communists woke up to it, it was too late. 'This was not just a "return to the roots", but also a sort of emigration from the normalised city, to the freedom offered by these small villages very far from the civilisation.'

On the Slovak–Polish border, in the tiny mountain hamlet of Brízgalky, Czech dissidents including Václav Havel would gather with their Slovak and Polish friends, far from the microphones and informers who haunted their steps in the capital, to drink beer and discuss things freely.

Before the fall of communism, they worked together with a wide range of people and unofficial groups, including the churches and Matica Slovenská, a cultural organisation dedicated to 'consolidating Slovak patriotism'. Under communism, anything that smelt of nationalism was banned and was therefore an ally for liberal-minded activists.

What of the return of capitalism, I ask. Mikuláš sighs into his beard:

The majority of us, in communist times, were not thinking about the return of capitalism at all, but of some sort of third way, incorporating the best of both systems. Instead, we have got what looks like the worst of both. It was extremely dangerous for a country which for forty-one years lived under a communist regime and had no experience of capitalism and democracy, to suddenly adopt a free market with all these negative features.

As land and businesses were privatised, new mafias sprang up, with money and influence left over from the communist era. The new entrepreneurs were the sons and daughters of the Communist Party bosses, and the common people missed out once again.

Before the Velvet Revolution of 1989, the Communist Party was responsible for everything. They had absolute power, but also absolute responsibility. But later, the responsibility was shifted to very different structures of local government, central government, the private sector . . . And now nobody's responsible for anything!

Mikuláš laughs sadly:

Local patriotism, meaning our sympathy and love for our native region, for the places of our childhood, is underestimated unfortunately by our liberal and democratic politicians and is tapped very effectively by our fascists and conservatives.

A gulf of sheer incomprehension has opened up between the village and the city.

CHAPTER 2

STONE WALLS AND FALCONS

A lamb sleeps like this in grass.
Slowly, with its own wool, darkness
covers it.

Milan Rúfus (1928–2009), Slovak poet and essayist

Beyond Devín, the Little Carpathians stretch away into the mist, bristling with castles overlooking the ancient amber route from the Adriatic coast to the Baltic Sea coast. The castles protected the northwest border of the Kingdom of Hungary.

I climb the steep path out of the village of Borinka with my youngest son, Jack, one January morning, from the rain up into the snow, among the grey trunks of beeches. The rain turns the leaves underfoot dark red, and the hard soil into slippery mud. A labyrinth of roots provides footholds. Here and there, tall fir trees interrupt the beech. Streams tumble down the hillside towards the Morava, or disappear mysteriously, to reappear on the other side of the mountains, near Limbach.

The top of Pajštún castle looms out of the mist, built on a rectangle of craggy limestone rock. This castle, lost in the forest, feels very different from the one in Devín, exposed on its promontory. Inside the ruins, the trees have been cut down or tidied up, and there's a fireplace and a simple wooden table.

Like Devín, Pajštún was blown up by Napoleon's troops in 1810, after he occupied Vienna, and was never rebuilt. According to legend, the mistress of the castle once turned away a beggar woman with two small children, jealous of her ability to bear children when she could not. The beggar woman cursed her to have octuplets and to suffer hugely in childbirth. The curse was fulfilled and the mistress of the castle bore eight sons. She handed all but one to a courtier, with the instruction to kill them. But the courtier took pity on them and brought them up as his own. When the son she had retained came of age, a party was held in his honour and his seven unknown siblings attended. Everything became clear, was forgiven and they all lived happily ever after.

On the far side of the Little Carpathians, above the wine-growing village of Svätý Jur is Biely Kameň, the White Stone castle. The road climbs steeply up a hill, past the church of Saint George. Then a track leads off into the woods to the left. It's still early morning in springtime and the forest is busy with birds. Dogs are out walking their people. The firm sunken track is narrow, hardened soil with oak and ash trees growing on both sides, their roots curling down the bank as the morning sunlight spills gold at my feet. In England, this would be called a 'lovers' lane'.

Just before the path reaches the ruins, an overgrown moat begins on the left. The crumbling white walls shine through curtains of bright green oak and beech leaves. At the top, a cluster of tall ash trees soars above the inner courtyard. On a rudimentary bench someone has carved a boy's name – 'Robert Patuška' – beside a big heart. Huge cumulus clouds billow in a dark-blue sky over the valley to the south-east. A thrush hops closer and watches me for a while in that fairytale way some birds have of appearing to be on the brink of human speech. There are signs of restoration, heaps of stone, sand and cement and scaffolding up against the walls. On a signboard there is a diagram of how the castle looked in the seventeenth century, with an impressive central tower.

There are fifteen wine cellars in the village. I taste a glass or two of the local Green Veltliner at the Vino Šmelko winery, which has a fine black-and-white print of an ancient vine laden with bunches of black grapes. Then we fall asleep in the comfortable guesthouse.

After the White Stone castle comes the Red Stone castle, or Červený Kameň. On my first visit here I met a young woman carrying a large falcon on a leather glove through the woods. She said a few words to it and it flew off high into the branches of a tree, perched briefly, then flew back to her waiting glove.

The falconry centre at Červený Kameň had more than seventy golden and steppe eagles, snowy and other owls, and several kinds of falcon in the large garden of an old stone building between the car park and the castle entrance. On my next visit, all the birds had disappeared; a sign on the gate mourned the death of Anton Moravčík, the man who had established the centre and overseen its work for many years. In 2023 it reopened.

The castle has the most remarkable underground cellars, where wine barrels, wheat harvests, silver, copper, gold and family heirlooms belonging to the Pálffy and Fugger families were stored. The Fuggers mined precious metals in a triangle of towns further east: Banská Štiavnica, Kremnica and Banská Bystrica. Nowadays the cellars are eerily empty, but you can see the magnificent arches and brickwork all the better. Upstairs there are two dozen rooms full of antique furniture, rugs and paintings.

Unlike most castles in the region, this one is still intact. There's a legend to explain the Pálffy coat of arms: a deer rearing up over a broken cart wheel. One evening, as a covered wagon with some members of the Pálffy family careered along a forest track, a giant stag leapt in front of it. The horses reared up abruptly and one of the large wooden wheels fractured. As it was nearly dark and too late to seek help, the occupants of the carriage made themselves as comfortable as they could for the night. In the morning, they saw that the track was about to plunge over a vertical cliff – the stag had saved their lives.

After the Morava river, the Váh cuts down the eastern side of the Little and White Carpathians, to reach the Danube at Komárno. At 400km and draining the High and Low Tatra mountains, this is Slovakia's longest river. In the late fourteenth and early fifteenth centuries, the Polish aristocrat Stibor of Stiboritz was known as 'the lord of the whole Váh' – only a slight exaggeration, as he owned fifteen of the thirty-one castles on its banks. The castle at Beckov on the Váh was his main home. I pass it on the motorway one April afternoon, the battlements square on its rocky promontory, the windows blind against the clouds.

These castles seem shrunk by the modern world, reduced from towering bastions of power to tourist trinkets. But the outcrops of rock on which they were built were chosen not just for their defensive possibilities, but also because they attracted local people, as out-of-the-ordinary places. Natural springs were discovered and rituals established to placate the gods of weather or else to win the favour of the gods of fertility or healing. People are still drawn here, to connect with something much bigger than themselves.

Ferdiš Duša was a Moravian artist who travelled down the Váh between 1927 and 1933, producing striking woodcuts of the castles and mountains on his journey. His woodcarving of the 'Black Váh' shows a shepherd in a broad hat, perched on a tree stump and clutching the small axe of his trade. His sheep are scattered around him and the snow-covered heights of the High Tatras rise up in the background.

The former lake is long gone and wheat is growing in its place. The artist would have been pleased to see the two church towers still in place, to the right of a castle. But the hills in the background again seem smaller now, without the romantic imagination.

◎ ◎ ◎

Ján Hacaj has a fine moustache and closely resembles the black-and-white portraits of his ancestors which line the walls of the wine cellar that his family rents from the Lutheran church. It's springtime, and we're standing in one of the biggest cellars in Pezinok, the next village along the edge of the Little Carpathians from Svätý Jur.

Ján was born in June 1960, which makes him four months younger than me, though in his company I feel like a child. There's a sparkle in his eyes and an understated smile. But also a seriousness, a quiet determination that has helped his family survive two world wars and the communist expropriation of their vineyards. His wife Ivana translates. As we talk, I often catch the names of grape varieties in the original Slovak before the translation arrives.

After the Velvet Revolution of 1989, he was a politician for three years, a member of parliament for Public Against Violence (VPN), the movement which helped topple the communists. He worked on the new land law, the return of land to people like himself, whose parents had lost it in the collectivisation waves of the 1950s.

Walking through the cellar, he describes the landscape above us. The gentle, southeasterly slopes of the Little Carpathians create an excellent microclimate for grapes.

No river runs along the foot of the hills, but there was once marshland. In the early morning, the first sunlight on the marshes raised a light mist, which in late summer would drift up the hillsides, through the vines, to settle on the grapes. This created 'noble rot', a shrivelling of some of the grapes into raisin-like shapes of almost unimaginable sweetness that were harvested after the first frost of autumn. Mixed judiciously with the pressed juice of the other grapes in carefully measured quantities, the result was very special. The Tokaj region, further east in Hungary, is the area best known for these grapes and for the aphrodisiac qualities of the wines they produce. Bottles from Tokaj are traditionally sent to royal couples in Europe as a wedding gift, to help expedite the arrival of an heir to the throne.

In the 1950s, the communist authorities in Slovakia drained the marshlands in the name of progress. The lowlands would be won for agriculture, and they were ignorant of the inevitable impact on the grapes above. A little autumn mist is left, Ján explains; but in 1990, when his family and their neighbours got their lands back, they had to experiment with other varieties of grape.

The communists also rearranged the direction of the vines, from vertical lines running up the hill to horizontal rows. The vertical lines favoured harvesting by hand, and were also fine for horses, which could go up and down the slopes quite easily. But horizontal lines were necessary to allow tractors to drive along the slopes. Such changes, and the roads built through the vineyards, made it impossible to claim back exactly the same plots as before.

While communist production focused on producing the maximum quantity of wine, after the Velvet Revolution the vintners could finally get back to the serious business of creating quality wines instead.

Ján's grandfather tended a couple of hectares. His parents were teachers and, like almost everyone in the town, split their time between their day jobs and the land.

Sparkling wine was one of the specialities of the Little Carpathians, and around twenty local vineyards produced it. It was also exported – a few bottles went down with the *Titanic* to the bottom of the North Atlantic in 1912. If the great ship is ever raised, the bottles might be retrieved, nicely chilled.

Ján's favourite place on his vineyard is where the Pinot Noir grapes grow today. In April 1939, his grandfather (also Ján) and his father, Vladimir, then aged thirteen, were planting vines there when the news came through that Czechoslovakia had been divided, with Slovakia becoming a puppet Nazi state under the leadership of a priest, Jozef Tiso. 'No point planting vines any more, son,' said his grandfather, laying down his spade. 'There's going to be a war.'

His grandfather was killed in the final days of the war, accidentally getting in the way of a bullet fired during one of the desperate battles that raged in these hills between the retreating Germans and the advancing Red Army. He had fought his way through the First World War in the Austro-Hungarian army and had then deserted to join the

Czechoslovak Legion, which fought its way across what was being turned into the Soviet Union. He only made it back to his precious vineyard in December 1920.

The freshly harvested grapes were poured through an opening from the street down into a big wooden winepress. From the high-ceilinged room we go down a few steps into the cellar. It's a cool and musty 8 degrees Celsius, while outside it's about twice that. The whole cellar is a masterpiece of brickwork – from the arched roof to the floor – with a symmetrical pattern of bricks interrupted by crazy patches, like a Turkish carpet. The bottles of sparkling wine are arranged in bands – yellow, dark green, yellow, like the flag of a new country – with temporary metal caps instead of the traditional thick, mushroom-like cork. Over four to six weeks, the bottles are turned by hand, at a precise angle within their wooden racks, allowing the yeast gradually to float up into the neck. At this point the necks – and only the necks – of the bottles are frozen. When the metal caps are removed, the bubbly elixir pops out the yeast, while the wine itself remains in the bottles. They can then be properly corked. All the work is done by Ján, Ivana, their son Matej and just one other permanent employee, though at harvest time they hire seasonal labour. Ján hopes his son will take over the business one day.

He leads the way upstairs to a room with comfortable benches, where Ivana brings small plates of hard cheeses, the darker ones smoked. Ján uncorks bottle after bottle for us to try. As a small sparkling wine producer in competition with the big wineries, which buy grapes from far and wide to sell in large quantities at a lower price, he has pondered long and hard how to keep his own bottles distinctive. His solution has been to make sparkling wine from unique grape varieties, rather than the mixed cuvées of the other producers. There's a Chardonnay, and two Romanian varieties of grape – Fetească Neagră and Fetească Albă (Black Maiden and White Maiden). He is especially proud of his Pinot Noir – a *blanc de noir* – white champagne from a black grape.

The word in Slovak for sparkling wine is *Šumivé* – which echoes the *sound* the bubbles make in the glass (whereas the English 'sparkling' describes what the wine looks like). Perhaps the Slovaks sip it with their eyes closed, the better to savour the flavour on the tongue, and listen to the fizzing.

Above the cellars, with the spring sunlight flooding the room, it tastes like the best champagne in the world.

CHAPTER 3

HERONS AND HEROINES

Since the birth of mass political movements, European nationalists have lamented the failure of their constituents to respond to the siren song of national awakening.

Tara Zahra, *Slavic Review*, 2010

In addition to one's so-called mother tongue, learned in childhood, one can learn another language, many other languages, and in that way step into other societies, be a member of other nations, as one wills . . .

Pál Hunfalvy (1810–91), Hungarian linguist

Much changed in Europe in the decades after Ľudovít Štúr's first excursion up the hill at Devín. National feeling fermented in the 1830s and 1840s and boiled over in 1848. The French Revolution of 1789, followed by Napoleon's revolutionary wars, spread ideas of national identity and civil equality. The monarchies and aristocracies of Europe were thrown onto the defensive. Romantic poets toyed with the new concepts, and moulded them into a different form in each country.

In February 1848, riots in Paris toppled the monarchy of Louis Philippe. The news spread on horseback across the continent and inspired uprisings against the rulers in Germany, the Netherlands, Denmark, Italy, Austria and Hungary. The national revolutions had a strong liberal streak. The rising middle classes of schoolteachers, lawyers and writers were out to break the monopoly on political power and end the feudal conditions in the countryside. The struggle was fought everywhere in the national colours.

Hungarians initially demanded more autonomy from Austria, then radicalised themselves to insist on complete independence. In the first military engagements they did well, with the support of several thousand Polish cavalry under General József Bem. But after a year of skirmishes, the Hungarian forces were overwhelmed by a Russian army sent by Tsar Nicholas I, who backed the Austrian emperor. They had a common interest in resisting the liberal tide, which threatened them both.

Hungarian resistance collapsed in the summer of 1849. Sándor Petőfi, the poet of the revolution, died on the battlefield at Segesvár (Sighișoara) in Transylvania in July 1849. His body was never found.

The national minorities of the Austro-Hungarian empire sided with Vienna against the Hungarians, in the hope of improving their own positions within the empire. This embittered the Hungarians against them, and helped sow the seeds for the emergence of new countries and the partition of Hungary in 1920.

The pan-Slav movement held its first, chaotic congress in Prague in 1848: some 340 Croats, Czechs, Dalmatians, Moravians, Poles, Ruthenians, Serbs, Silesians, Slovaks and Slovenes and 500 guests attempted to find common positions amidst the thrall of their own national feelings, caught between the movement for the unification of Germany on one side, and the growing nationalism of the Hungarians on the other. The Czech liberal František Palacký, a representative of 'Austro-slavism', was president of the congress. His idea was to have 'a transformation of the Habsburg monarchy into a federal state, in which Slav nations would give up the idea of full political independence in favour of cultural freedom within Austria'.

On 12 June, Austrian soldiers in Prague opened fire on peaceful demonstrators and the congress broke up in disarray. Its main achievement was a 'Manifesto to the Nations of Europe', to be delivered to the Austrian emperor, calling for an end to the oppression of the Slavs, and the transformation of the empire into a federation of equal nations.

Back in Buda and Pressburg (Pozsony, Bratislava), the Hungarian literary elite were also good students of Herder and his ideas of an 'organic' nation. They were haunted by his remark that '. . . in a few centuries their language may no longer be found'.

Herder noted that the Slavs and other nationalities were numerically superior to the Hungarians. The 15 million Hungarians in the Carpathian basin in the mid-nineteenth century were outnumbered by some 20 million Romanians, Slovaks, Germans, Serbs, Croats, Ruthenes, Jews and Gypsies. In 1910, Hungarians comprised just 55 per cent of the population of the lands they administered.

The efforts of the romantic nationalists to strengthen the sense of national belonging were complicated by the fact that ethnic distinctions were blurred by religious affiliations and inter-marriage. Hungarian and Slovak Roman Catholics worshipped side by side, as did Hungarian and Ruthene Greek Catholics. Priests and pastors expected loyalty to the true faith and were wary of the growing faith in the nation.

Two different concepts of the nation competed in the mid-nineteenth century. The first identified the nation with language. True Hungarians could only be those who spoke Hungarian as their mother tongue. Knowledge of other languages was frowned upon as somehow diluting one's 'Hungarianness'. This was also true of the other emerging nationalisms. The linguistic philosophers and romantics set out to define what made them different from other peoples, rather than what they had in common. The idea that a nation can be 'sovereign' vied with the idea that an individual is sovereign, in front of God.

'Our first task is to reveal the inner secrets of our own language,' wrote Gergely Czuczor, co-author of the first etymological dictionary of Hungarian (1862–74). Just as the Hungarian lands needed to be liberated from Austrian rule, so the Hungarian language should be 'liberated' from Latin and German words. The language should be centralised, standardised, and local variants eradicated. The multilingual and multicultural Austro-Hungarian empire, where Latin was still the language of politics and German the language of administration, was seen as an obstacle to the sense of national identity of the people.

The rival concept, championed by the linguist Pál Hunfalvy, argued that a Slovak or a Serb, a German or a Jew would be more patriotic to Hungary if they were allowed to speak their own language: 'In addition to one's so-called mother tongue, learned in childhood, one can learn another language, many other languages, and in that way step into other societies, be a member of other nations, as one wills.'

This was a generous, broad-shouldered approach, and one which many families practised, for the common-sense reason of wanting to improve their children's prospects. Temporary 'child exchanges' flourished in the late nineteenth and twentieth centuries. Children aged between eight and twelve were sent for periods of up to a year to stay with another family in an area with a different linguistic majority, to learn their language – Hungarian, Slovak or German. They also attended the local school. Long-standing friendships between families, ethnic tolerance and even marriages resulted. The practice was popular among the middle and lower nobility, farmers, lawyers, merchants and other professions, while landless peasants sometimes sent their children away as household servants, to learn another language.

The 1868 Hungarian Nationalities Law guaranteed the right of minorities to use their own language. But over the following forty years the monolinguists, armed with the new tool of the national census, suppressed multilingualism and succeeded in further alienating the nationalities with their efforts to 'Magyarise' them.

◎ ◎ ◎

The River Hron rises in the Low Tatras and flows 270km to the Danube at Štúrovo. The river basin arches neatly under that of the Váh, like a baby in its mother's lap. Until the Second World War, spruce trees felled in the valleys of the Low Tatras were transported downriver on the Hron, tied together in rafts. At Zvolen, where the Slatina flows into the Hron, four mountain ranges meet, all rich in precious metals: gold from Kremnica, silver from Banská Štiavnica and copper from Banská Bystrica. A third of the gold in Europe in the Middle Ages came from the Carpathians – either from these mountains or from those in Transylvania.

Three heron nests nestle high in the poplar trees overlooking the car park of Zvolen castle. *Volavka popolavá* is rather poetic in Slovak: *popol* means ash, so these are 'ashen' herons. Beneath the nests I meet ecologist Marta Pauliková and ornithologist Anton Krištín. There are four eggs in one of the nests, Anton explains. All over Europe herons, like many other wild creatures, are moving into the towns, as if by common agreement – you've been taking over our space for so long, now we're going to take over yours.

The castle beside us is a solid rectangle of stone, bruised but not broken by the waves of intrigue and invaders that have swept through the foothills of the Carpathians. An older ruin, known as the 'deserted castle' – *pustý hrad* – perches on a forested hill on the other side of the Hron river, overlooking the trading road from Buda to Kraków. We drive out of town on the main road beside the Hron, then branch up a narrow track beside the river. It is mid-May and all the trees are in leaf, washed by the wind and rain of the previous evening. *Spomaľte, žaby!* read the signs beside the track: 'Go slow, frogs!' The water is muddy from the rains.

'Among birds, but not among humans, the males are the more beautiful, and the grey heron is a good example,' Anton tells me with a twinkle in his eye, as we trudge beside the Slatina. He moved here from Bratislava. His wife did not enjoy the life of the provincial town, and left for Prague. His two sons grew up and moved to California. 'I am here as a bat,' he says, in somewhat bat-like English, and I guess he means perched in a tree (not necessarily upside down), flying around but always coming back to the same place. Each year he spends two months travelling for his ornithological studies, and is just back from the island of Réunion in the Indian Ocean.

'I am happy here, equidistant from Budapest, Vienna, Prague and Kraków. I have my mountains, my birds, my brown bears and lynxes. And all the mineral springs . . .' The region is as rich in hot and cold water springs as it once was in precious metals. 'These volcanic hills are full of energy,' Anton says, and he is not talking about hydroelectric power.

Marta grew up in Svätý Jur, studied ecology and environmental sciences at the University of Zvolen, and fell in love with the rivers of central Slovakia. As a student in the early 1990s, she was brought here by her professors, to see the rich flora and fauna of the Slatina river basin which, they were told, would soon be lost forever beneath the storage lake of the hydroelectric project. The river meanders beautifully for 50km from its source in the Poľana mountains. Marta and her fellow students decided to save the river: they formed the Slatinka Association.

'We thought we might have to campaign for six months,' says Marta. 'We had no idea that we had a thirty-year fight on our hands!' The engineering lobby in Slovakia is formidable. Rooted in a communist era when nature was seen as a prone beast, waiting to be exploited for the greater glory and comfort of man, Slovak engineers built twenty-nine hydroelectric projects on almost all the country's large (and smaller) rivers. The biggest – the Čierny Váh on the Váh river near Liptovský Mikuláš – flooded three villages forever under a broad storage lake, 45 metres deep.

'The reasons given for the planned dams on the Slatina changed with time,' she says, as we wade gingerly through marshland bristling with horsetail, also known as puzzle grass. 'At first they said they needed the water for hydroelectricity. Then that they needed it as coolant for the Mochovce nuclear power plant, 80km away. Finally they said they needed the water because of climate change. But if they had really needed it, I'm sure they would have built the dams, whatever we said.'

In 2013, Slovakia asked the European Union for money to build the dams, but the request was rejected because of the environmental damage it would have caused.

The marsh horsetail, *Equisetum palustre*, bristles and rustles in the breeze. It is one of the most ancient grasses on earth, dating from the Jurassic era 100 million years ago, the time of the dinosaurs; and it is said to have inspired John Napier, the Scottish mathematician, to invent logarithms in the sixteenth century.

The dam project, paradoxically, saved the riverside from development. As the river valley was slated for destruction, no building was allowed along its shores. To stop the dams, Marta and her friends organised petitions, leaflets, marches and study tours. Zuzana Čaputová, president of Slovakia in 2019–24, worked for the association as its lawyer.

There are delicate purple flowers in the undergrowth, guinea-hen flowers or leper lilies, *Fritillaria meleagris*, so named because the spots on the purple resemble those on a guinea-hen, and the flowers the bells that lepers had to wear around their necks to warn people off. It was once so common in England that it was sold in bunches on the streets. The ploughing up of ancient meadows to create more farming land during the Second World War is blamed for its disappearance.

In the poplars at the water's edge are boxes for bats – the lesser noctule, *Nyctalus leisleri*, which live in trees, fly extremely fast and look like lions due to the thick golden-brown hair on their shoulders. It's wet underfoot and we tread gingerly through a meadow of bright yellow marsh marigold, *Caltha palustris*. The land is like a sponge, giving and absorbing water, Marta explains. This is what sustains biodiversity. When we drain it, all this is lost.

The first bird we hear singing is the Eurasian blackcap, *Sylvia atricapilla*. Anton produces from his backpack a small audio recorder and a loudspeaker, and plays the call of the male kingfisher. The sound of insects and the individual calls of the birds untangle themselves in the soundscape. A flash of blue, then it's gone – another male, curious about this peculiar intruder on its territory.

The kingfishers, black caps, kestrels, eagles and owls were allies in the campaign to save the river. The ecologists didn't confront the dam-builders: they pretended to go along with them, but set many conditions:

We demanded at least three islands in the storage lake, planted with alders. Cliffs where birds like the kingfishers and bee-eaters can nest, and a special habitat for the two pairs of lesser-spotted eagle, *Clanga pomarina*, the smallest eagle in Europe, which also breeds in this river valley. We made it so expensive for the energy firms that in the end they gave up.

The kingfishers breed in the dried mud walls on the banks of the river. Anton spots a collared flycatcher, so-called after the white band around its neck, nesting under the roots of the alder trees on the bank. We study it through his binoculars. Then he points out a nuthatch, a small bird with a disproportionately large head and strong beak, with a habit of wedging nuts or other food in the fork of a branch, in order to hack at it.

The path takes us upstream, with the river on our right, through bushes of white buckthorn. In traditional medicine, this is used as a mild laxative, for digestive problems, and to heal sore throats. There's a sudden bright yellow flash in the trees to my right, probably a golden oriole. Anton imitates its call to tempt it back.

The river meanders between stands of alder trees. There are the white droppings of otters on the rocks in mid-stream, then our first grey heron, standing completely still: a male, judging by the distinctive black stripe close to the eye and the long, ornamental feathers on the back of the head. The dignified silence of the heron contrasts with the busy song of the mistle thrush, the distant cuckoos and the constant hum of the crickets from the meadows nearby. Anton points out a white-throated dipper, *Cinclus cinclus*.

This section of the river is also a favourite feeding ground for lesser spotted eagles. 'They eat mice, lizards, snails, even rats. Occasionally dead fish if they find them on the banks. They are like the vultures you find in the tropics!'

The trees sprout from a complex root system, up to twelve trunks from the same roots. These are the alder 'carrs' – a waterlogged wooded terrain populated with alders, for which the Slatina is famous. The peculiarity of alder wood is that when it is submerged in water, even salt water, it does not rot but becomes hard as stone. The city of Venice was built on alder piles.

'These trees are very old. A hundred, perhaps two hundred years old at the base.' The river spreads so wide, it is hard to discern if we are on the mainland or an island. There are also patches of almost still water, excellent breeding grounds for fish and frogs and small amphibians. One is the fire salamander, *Salamandra salamandra*, a strange black, lizard-like creature with yellow spots.

A mountain wagtail next, a small grey bird with a white chest and distinctive black-and-white feathers and tail. The water level in the river is low; but in August, when the waters gather in the mountains, it can reach 2 metres deep. We pass various sedges, and strange ferns like the tails of ostriches. The rivers of central Slovakia were once all like this, a sub-montane landscape, before they were so extensively dammed. Now it is very rare. 'The water here is not only from the river. It's coming up from the ground!'

The white droppings of a black stork, another rare species, mark the stones: these are fresh droppings, from this morning, after the rain stopped. Two or three pairs of black storks, a shyer bird than the white stork, live on this stretch of the river.

'So-called normal people spend their weekends in shopping malls, and never go out into nature. Most people don't realise the power of nature, because they have never seen a really living river.' Anton stoops to catch an agile frog, *Rana dalmatina*, almost invisible in its grey-green camouflage. It is 'agile' because its powerful back legs are unusually long for a frog, allowing it to leap up to 2 metres. Anton lays his frog gently down on a stone and it leaps spectacularly away, in amazement.

'Probably his first encounter with a human,' he says. Four mallard ducks fly high overhead.

The river gets faster, unexpectedly. A yellow wagtail starts singing – our second kind of wagtail in the space of ten minutes. Marta shows me a deep pool where she and her family swim in summer. We turn back through the forest, past magnificent oaks, including a pubescent oak, *Quercus pubescens*, at least 300 years old. Each bulge an eye, each welt in the bark the muscle of cheek or curl of eyebrow. It would be a strange place to walk at dusk. The trees watch us, moving slightly – imperceptibly – apart to let us pass.

Slovakia has a complicated ownership structure, still marked by traces of the feudal system, the collectivisation that occurred under communism and the multiple impacts of capitalism. This particular forest belongs to the *urbariát*, the local community, a system of collective ownership dating back to the end of feudalism in 1853.

The *urbar* was a kind of charter in the Middle Ages, laying down the concessions that the aristocratic owners of the land made to the peasants who worked it, and what the latter had to do in exchange, like cutting wood and transporting it. For the ecologists, such collective ownership forms can present a significant problem, if those gathered in the *urbar* decide to chop down the oaks and sell the valuable wood.

Anton's next find is a tiny grasshopper which lives in deadwood, when old trees are not cleared away from the landscape. It can even live underwater.

Emerging from the woods into a meadow, we search in vain for the purple marsh orchid in the deep rustling grass, two weeks too early. The meadows are mown by tractor each year, but there are plans to reintroduce horses.

As we reach the car, a lesser kestrel, *Falco naumanni*, circles overhead. Then Anton hears a wryneck – the only species of the woodpecker family found in Europe that migrates. It takes forty days to fly 10,000km from Botswana or Namibia. In the autumn, the birds make a more leisurely journey, about ninety days, pausing more often to rest and feed. Anton installs his loudspeaker at the foot of a hedge, and sends out the long, shrill call of the male wryneck. Almost immediately, there are two answering calls, then another. 'A male and a female!' There's a flurry in the air, and the male perches in a small tree, close to the hedge.

CHAPTER 4

IN PRAISE OF GOATS

Tears of happiness were coursing down my cheeks. I was running forwards, not backwards.

Rudolf Vrba (1924–2006), escapee from Auschwitz

The ancient prison perches near the centre of the Hungarian town of Balassagyarmat like a mole on a wrinkled hand. Built in 1840, its metre-thick sandstone walls still present a formidable obstacle to several hundred inmates.

Balassagyarmat sits beside the Ipoly river in northwestern Hungary, close to the border. The Ipoly rises in the Ore mountains in Slovakia as the Ipel' and flows all the way down through the Börzsöny hills to the Danube bend at Szob. It seems particularly cruel that the cell windows of the jail all look inwards, rather than outwards to the Carpathians. In the prison chapel, the roof beams are of pine, cut from the slopes of the Low Tatras.

We drive north through Balassagyarmat in early September under blue skies, the hills woolly with trees. The fields at the roadside are still emerald green, the hillsides alternate dark green with spruce, lighter green with beech; but the scent of autumn is in the air. We're bound for Opavské Lazy, just across the border in Slovakia.

Laco Bakay waves us into his garden. A horticulturalist with a thin beard and big smile, he specialises in rare trees at the University of Nitra. It's one of the family gatherings with which they start and end the summer – a big cauldron of goulash bubbling on an open fire. As well as Laco there are his wife, Libuša, their seven-year-old son and fifteen-month-old daughter, and Libuša's parents from Poprad in the east of the country. We eat fresh blueberry cake washed down with coffee. Then Laco takes us on a grand tour of his trees, while mother and baby have an afternoon nap.

The oddly named service tree, *Sorbus torminalis*, is the oak tree of the fruit trees: it can live to 300 years and grow 30 metres tall. The Latin name *torminalis* means 'good for colic', and in the medical herbals of the Middle Ages its value for healing the stomach pains of humans and beasts is legendary. The fruits are best when they seem over-ripe – they are very soft and are full of seeds, which need to be judiciously spat out. But the taste is divine, like stewed apples with blackberries. Laco has many different varieties of apples and pears, hawthorns and sandthorns, sweet chestnuts and walnuts. His whole garden is an experimental orchard of strange and wonderful varieties.

This area was famous for its cherries, and there are the remnants of long avenues of cherry trees looping over the horizon. Seven different varieties keep the harvest going over a six-week period. They used to be taken by horse and cart to Vác on the Danube, loaded onto ships and taken downriver to the hungry markets of Budapest.

Laco describes drinking champagne here with his wife, up to their waists in thick fog, unable to see where they were treading. Today it's a beautiful Indian-summer afternoon, small white clouds dotting an otherwise blue sky, volcano-shaped hills along the horizon.

We visit a neighbour's farm full of tame goats who follow us through a plum and apple orchard. Brown goats with small horns and tiny purple tongues run bleating, the kids playful but fearful, the adults more challenging. One climbs up into an ancient pear tree. The goats recognise Laco, and greedily accept bunches of leaves from him.

In such an orchard, you have to decide which fruit trees you leave to the goats, and which you harvest yourself – to eat or turn into *pálinka*. Cherry trees live for sixty or seventy years, and one of Laco's plans is to plant new ones. They need a lot of water, which means a lot of care when they are young. The older trees need careful pruning, as branches that get overloaded with fruit can easily crack. If you reduce the crown, the roots dig deeper into the earth, strengthening the tree.

There are cowpats in the grass, from which perfect white mushrooms sprout. An ancient walnut tree bends right down to the well, as though intent on taking a sip in the human way – from a lowered bucket – rather than drawing the precious liquid up through its roots. Laco introduces us to Meki, a smallish white goat with a unique bleat of his own, something between a cockerel and a crow. Laco makes the same cry back, and for a while they exchange remarks – about the weather, perhaps, or the taste of the leaves. When children come here, he says, they invariably imitate Meki. Laco's son will soon start primary school. 'Perhaps he will make the Meki sound in an arithmetic class to confuse the teacher!'

It's only two hours' drive from Lazy to Nemecká nad Hronom – 'the German village on the Hron' – in the Low Tatras. Mišo 'the Fish' Saradin runs rafting tours from here, down the Hron. We pitch our tent on the river shore, beside another tent and a fireplace. Soon our neighbours return – a psychiatrist, a police liaison officer and their children – and we spend a pleasant evening chatting by the fire, sharing food and wine. We fall asleep to the sound of the shallow Hron running over the stones, and the crackling of the fire. My first night for a while under canvas, and I sleep badly, woken by every tread or bark or murmur.

Mišo was born 'thirty metres from the river', he says, and knows every twist and turn of the Hron. In the morning, he takes us out for a paddle. 'When the water is low like now, the bigger fish take shelter in the deeper hollows.' Swimming beneath us close to the bank, where the water level reaches 1.5 metres, are huchen, *Hucho hucho*, also known as

Danube salmon, which are hardly to be found in the Upper Danube anymore because the water is too warm for them. The remaining fish shelter further and further up the mountain rivers. They need oxygen and a strong current. On sunny days they sometimes jump; but today it is overcast and they stay out of sight. A great egret flies up ahead of us, from its perch on a sand bank, then settles each time just upstream, as though showing us the way. The white-throated dippers are more frequent, flitting from stone to stone.

There was no rain at all this June and July, and only four rainy days in the whole of August, so the Hron is at least 25cm lower than usual, which forces us to paddle our Kevlar-bottomed raft fiercely, to stay afloat. What in normal times is an island appears ahead of us – and we steer carefully to the right, always seeking the deeper water. Mišo fishes here, but always puts the huchen back. 'In this hollow, the water is still 2 metres. About twenty huchen live here. Last winter, I caught ten and took pictures of them. All of them were different.' He can tell the age of the fish by the number of spots under the eyes. In this particular hollow, ten years earlier, the king of the huchen used to live – 1.3 metres long and weighing 22 kilos. 'In 2021, with our group of rafters in the local anglers' club, we offered seventy special permits for catching huchen on this stretch of the river. Only seventeen were caught, and all were put back.' Mišo seems to know all the fish in this river so well, he might almost give them names. 'We reckon we should keep about one in five of those we catch, as the huchen is a predator for our brown trout, *Salmo trutta*, grayling and char.'

We pass a place on the right, just before the village of Brusno, where the tracks of animals that come down to drink are clearly visible in the mud. Mišo sees deer and wild boar regularly here, but has also seen bears. The village is famous for its nine healing springs. Beyond that, at one turn in the river, there are outcrops of flat, grey, slate-like rocks, near a place where he once caught a 7kg brown trout. There are even river crabs. There's a turquoise flash – a kingfisher. The Sopotnica stream brings a trickle of water to the thirsty Hron, and we look up to its source, between Veľká Chochuľa and Košarisko in the Low Tatras, two peaks on the hundred-kilometre ridge walk from Donovaly to Kráľova hoľa.

The church bell in Brusno strikes one as we pass the ruins of a bridge, blown up in the Second World War, and paddle under its replacement, our voices turning deeper with the echo. This was partisan territory in 1944, and Mišo's grandfather used to take food up to the fighters in the mountains that autumn. There were only two areas that the German troops failed to take back from the partisans, he says, and one was the Upper Hron basin. The Germans were forced to put up 'Achtung Partisanen!' signs in the forests around here.

A train roars out of nowhere, screeching over the railway bridge and deafening us for a moment. There is less plastic rubbish in the Hron than there used to be, Mišo reckons – partly thanks to regular cleanups and partly due to a gradually increasing awareness among members of the public that it is not acceptable for them to leave rubbish behind, when they picnic on the shores.

That afternoon we drive in a white Land Rover with a rubber snake-head fitted on the gearstick, up the mountain through Jasenie with Mišo's friend Brane, a ranger in the national park. On the way, we stop to examine a log hut. The tarred logs have been scratched away in several places by inquisitive bears. There are more bears than there were five years ago, but they rarely cause trouble – only if people behave stupidly in their vicinity. 'We locals are OK with the bears,' he says. A big man in a green T-shirt and baseball cap, Brane stands up against the cabin with his arms in the air, to give us an impression of the size of the bear that did this.

Mišo has a shaved head, green eyes, wrap-around shades and a camouflage jacket. He makes a living in summer taking people up and down the river. As well as having been a fisherman, rafter and hunter, he has worked for the Slovak mountain rescue service. It once took him eight hours, in the deep mid-winter snow, to reach the Ďurková mountain cabin above us on the ridge. This mountainside is notorious for its avalanches, when the snow starts to melt in spring.

To the left of the rough forest track, we stop again to look at a partisan monument – a grave with a square concrete base and green iron railing. The stone has an army helmet on top. Many partisans came from the ranks of the Slovak army. They occupied a large swathe of central Slovakia, before they were pushed back up into the mountains by the German army and resorted to guerrilla warfare from bases like this one.

Here, far from his dearest ones, lies our only son Karol Sládek, a partisan commissar from Bratislava, who died a heroic death on 30.XI.1944. His comrades-in-arms, the partisan Hudec from Ľubietová, and the partisan Veselovsky from Hostetno, lie beside him. In loving memory.

There was a partisan field hospital in the forest close by. Two thousand people lived on this mountainside in the autumn and winter of 1944. Now, briars climb over their tombstones, loaded with fat, ripe blackberries. On Karol Sládek's grave there are fresh red and white roses and wreaths with tricolour ribbons.

There's a sign in Slovak and English with a picture of a large mother bear followed by two cubs, with some useful advice: 'Territory of Brown Bear. In a rare chance of meeting, avoid direct eye contact, back away quietly. Stay calm, avoid sudden movements and gestures. Keep away from cubs, mother is definitely nearby.'

Above the treeline, a carpet of blueberries and cranberries stretches either side of us, autumn turning the leaves from green to gold. Below them is a dark-green expanse of spruce. The Ďurková shelter has a modern feeling from the outside – a wooden house built on a stone base, with walls of pine logs, large windows at the front, and a steeply sloping roof to combat the snow. A line of colourful Tibetan flags flutters in a sharp wind. It was built in 1932 on the foundations of a sheep shelter and was burnt down by the Nazis in the war; it was only fully restored in 2000.

There are photographs on the walls of rosy-faced hikers, and records of heroic deeds. Several young people arrive with rucksacks on the veranda outside, where we sip bowls of vegetable soup. Martin and Tomáš are students of electrical engineering from Brno, and have walked for ten hours to reach here from Donovaly. Then a Belgian couple hurry in, excited by a sighting of a mother bear and her cubs, just a few hundred metres away from the cabin. There are tents pitched close to the cabin. Inside, about a dozen people are playing cards or resting.

Rain spatters our faces and smudges the ink in my notebook. Row after row of mountains, right the way back to the border with Hungary. On the clearest days, two or three times a year Mišo says, you can make out Hoverla, the highest mountain in the Transcarpathian range in Ukraine, 170km to the east.

◎ ◎ ◎

In the valley of the Hron in Banská Bystrica, the museum of the Slovak National Uprising is a cross between a spaceship and a pair of sweet chestnuts, cupped together in a concrete shell that has been peeled away. There's a remarkable collection of guns and maps and parachutes, but few individual stories.

In April 1944, Rudolf Vrba escaped from the Auschwitz concentration camp with his companion, Alfréd Wetzler, to alert the world to the mass murder of European Jewry. Vrba, aged just nineteen, and Wetzler made their way southwards, through Nazi-occupied Poland and across the Carpathians, carrying with them diagrams, maps and details of the numbers and methods used at Auschwitz. They crossed the Polish–Slovak border at Skalité, near Zwardoń. In a safe house in Žilina, they wrote up a thirty-three-page report. Oscar Krasniansky of the Slovak Jewish Council translated it into German. It was sent to governments, the papal nuncio and the media. The first extracts appeared in the *New York Times* and on the BBC Overseas Service in June 1944.

At the Svätý Jur monastery near Bratislava, Vrba and Krasniansky met the Swiss legate Mario Martilotti on 20 June. He took the report by train to Switzerland and cabled the contents to Pope Pius XII. The pope asked the Hungarian regent Miklós Horthy on 25 June to 'do everything in his power to save as many unfortunate people from further pain and sorrow'. Horthy's wife noted in her diary that she went with her husband to meet the papal nuncio Angelo Rotta in Budapest on 3 July. On 7 July, Horthy, who had only limited power after the Germans occupied Hungary at the end of March, ordered an end to the transports. The dates are important, because each day 12,000 Jews were packed into cattle wagons in Hungary in appalling conditions and sent to Auschwitz. Many died on the way. Ninety per cent of those who reached Auschwitz, according to Vrba and Wetzler's notes, were sent straight to the gas chambers, on arrival.

Western leaders had known about the extermination camps for years, but not about the sheer scale and horror. That was the contribution of Vrba and Wetzler; but their accounts leaked out painfully slowly. Around 120,000 Hungarian Jews survived the Holocaust, and the report undoubtedly played an important part. But more than half a million perished between the day the two men escaped and the day the transports stopped. Vrba claimed to the end of his life that his report was suppressed, in part by the British government (officials told journalists they thought it was exaggerated) and in part by Slovak and Hungarian Jewish leaders, afraid it might disturb their negotiations with the Nazis for the safe passage of Jews.

The Slovak National Uprising began on 29 August 1944, the day German troops occupied Slovakia. Around 48,000 took up arms and managed briefly to take control of central Slovakia. The Germans threw some of their best remaining troops into the battle. By mid-October, most of the partisans were on the run, dead or captured. Their strategy shifted from open confrontation to guerrilla resistance from their camps in the high mountains.

Stalin ordered the start of the East Carpathian offensive, to break through the mountains at the heavily fortified Dukla pass on the border between Poland and Slovakia. Dukla is the lowest pass in the Carpathians, at just 502 metres. Marshal Konev, the Soviet commander, hoped to cross it and reach the eastern Slovak city of Prešov in five days. It took him fifty just to reach Svidník, a small town just the other side of the pass. Some 70,000 men were killed or injured on both sides, in one of the bloodiest encounters of the whole war. The landscape around Dukla is still a war-scarred, mournful place.

◎ ◎ ◎

The Low Tatras run 90km along the southern edge of the Váh river valley, some 30km south of their big sisters, the High Tatras. On my green-and-brown relief map, the two ranges are like twin folds on a crumpled blanket.

In a meadow just east of Telgárt, a lone cat guards the source of the Hron river – a thin trickle of sweet water. There are little log structures with wood-shingled roofs, just enough to shelter in on rainy days. Above, the bald beauty of Kráľova hoľa, the King's Mountain, the last and tallest of the long knuckled ridge of the Low Tatras. On the map of Slovakia, it is slightly off centre; but because of the way the whole country tilts westwards towards the capital, this feels like the beginning of the real East, closer to Ukraine and Russia, the western fringes of Asia.

Kráľova hoľa is the birthplace of four rivers. The Hnilec flows, small and chatty, beside us as we climb the mountain, on its way to the Hornád, which becomes the Hernád when it crosses the border into Hungary, to reach the Tisza. Meanwhile, the Black Váh flows northwards to merge with the White Váh (which comes down from the High Tatras) and become the Váh, the Hron's great protective wing.

The going is easy on a sunny October morning. We cross the stream over a small bridge and strike out uphill, through thick forest; then comes a hillside stripped bare by logging, and then a forest of spruce, laid low by the wind. The hillside at Tri Mosty, 1,100 metres, is devastated, just tree stumps left in the long yellow grass of autumn, only one or two spindly trees remaining standing. At one point, the path winds along the edge of a reservation, with tall pines and beech. Then we come out into the open, with dwarf juniper, *Juniperus sibirica*; the peak is straight ahead, up the steepest incline.

A TV transmitter, 138 metres high, towers over an ugly building; but nothing can spoil the feeling of happiness at having reached the top of the mountain, with a chance to scan the horizon in all directions.

The King's Mountain was named after Hungarian King Matthias, who cut his name into the rocks of the King's Steps, a little further down the slope towards Telgárt. For Slovaks, the mountain is one of many hideouts for their hero, the highwayman Juraj Jánošík (1688–1713). According to legend, he was captured in an inn, after slipping on peas thrown onto the floor by an enemy, and was executed in Liptovský Mikuláš, aged twenty-five. He is remembered in song and legend:

> We'll go, Mama, to the Slovak forests,
> to that world of beauty and passionate youth!
> There the grey fir is waiting for us,
> stretching out her arms from afar:
> 'Welcome, children!' She invites us to join her,
> 'I will be your mother!'

◎ ◎ ◎

On a moonless, starlit night I walk with Samo Hríbik up a steep meadow in his village. A shooting star sweeps across the sky in front of us, like a gift to those still out in the fields, when others are snug inside. We're going to gather his sheep and goats and herd them into the stable. A few years earlier, one July night, wolves killed seventeen of his sheep and lambs. 'I got so angry, I called the hunters. I wanted revenge,' he says. Now his anger has gone and he regrets the decision.

It's hard to find twenty-one sheep in the dark. We decide not to use torches, so that our eyes become accustomed to the dark. 'It's in the

nature of sheep to climb. Probably because they were originally mountain animals . . .' He let them out this morning. 'Sometimes they stay
close to the stable, but mostly they climb . . .'

We go from one meadow to the next, leaving the lights of the village
below us in the valley. He calls out to them, a string of sounds he learnt
from his uncle. 'They get used to it, because I repeat it, every evening of
their lives.'

Suddenly he freezes: 'There, hear them?' I hear nothing. His ears are
finely attuned. We strike out across the next field. Finally, I hear them
too, a gentle tinkling from close by; then we see their shapes, clumped
together, blocking out the lights of the village. Every fifth sheep has a
bell, and the bells chatter to each other. Then one or two sheep bleat a
greeting in return.

We loop around the top of the field, and they start moving, a slightly
lighter cloud or shape in the darkness, down the hill towards the stable.
They're surer on their feet than me, but not than Samo. He guides them
down the hill with his voice. Unlike other shepherds, he has no dogs.
'By the time you've paid for the dog food, there's not much profit in
sheep. There's not much profit in them anyway,' he shrugs. He breeds
them for their meat. The ewes give birth in February and he sells the
lambs in August. Until then, they have a good life here in the hills.

He also has a few goats. 'For a kilo of goat's cheese you need 12 to
15 litres of goat's milk – much more than you need to make cheese
from cow's milk.' He doesn't milk them, because there is no tradition of
consuming goat's milk or cheese locally. 'What I like about goats is that
they are so unpredictable. And they eat the shrubs and the bark, and the
leaves of the trees. So they slowly restore the former meadows and
pastures.'

Samo also has five beehives, replacements given him by a friend after
his own bees died last winter. The bees find pollen at this time of year
from the highly invasive goldenrod, *Solidago canadensis*, first introduced
to eastern Europe from North America in the 1870s. Beekeepers favour
it because it continues to flower in September and October. The

downside is that it inhibits the growth of other flowers earlier in the summer, and is harmful to wild bees, butterflies and ants.

His grandfather hung on to some land, even under the communists, but had to give most of his produce to the agricultural collective, as punishment for being a private farmer.

My vision is to make 'pasture forests'. That means half forest, half meadow. Because this is the ecosystem that is the most rich for biodiversity. But it's very rare in Slovakia, because no one pays the farmers a subsidy to create that kind of landscape.

There are 1,400 inhabitants in Telgárt, including a good number of Roma, in poor housing in the lower end of the village. The Roma are better off here than in many places in eastern Slovakia, says Samo, because there is less abuse of toluene – a kind of paint thinner – here than in other Roma neighbourhoods.

Some locals commute to Poprad for work, 30km away. Samo feels privileged because his parents own land. They started to buy after the Velvet Revolution of 1989, and now have 15 hectares. Samo studied ecology, and has worked the family land since he graduated. In winter, he carves flutes and wooden tools.

The next morning after breakfast Samo shows us his workshop. The walls are lined with rakes, made from three kinds of wood: the teeth are of ash; the frame is of beech; and the handle is either hazel or spruce. The handles are unusually long so that the can rake up the hay on the steep hillsides. He was inspired to make them by his great-grandfather, born in 1901, who built the house where Samo lives. He inherited his chisels and his workbench. Samo also collects old axes and axe heads. 'Some people just trade them for bottles of wine.'

The flutes are fashioned from elder. He plays a very long one, a *fujara* flute from central Slovakia, in long, shuddering notes, like the wind buffeting a thatched roof. Smoother, sweeter, more melodious sounds emanate from underneath. It has a sort of double cylinder at the top, with a protruding wooden mouthpiece into which the player blows, reaching for the holes at waist height.

Samo's brother Matúš is a forester working just outside Telgárt in Pusté Pole. 'The whole forestry in Slovakia is in big crisis,' he says. The industrial loggers have the upper hand. 'Nobody wants to live here permanently. They just want to cut a lot of trees, make a lot of money, and retire.'

On 19 November 2004, a northerly wind picked up over Poland, crossed the ridges of the High Tatras, then hammered down on the Slovak forests, reaching speeds of up to 170km per hour. The storm ripped through a wide swathe of the High and Low Tatras, flattening almost all the trees in its path. Worst affected were the large plantations of closely planted spruce, which fell like dominos. The more spaced-out mixed forests, which develop a deeper and more complex root system, were also badly damaged, but not quite as badly.

For the forestry industry, the storm was an unmitigated disaster. For the ecologists, it was an opportunity to begin again – and let the forests regenerate naturally. At the centre of their quarrel is the bark beetle, *Ips typographus*.

After such a storm, known as a 'windthrow', the traditional response of the forestry industry is to extract fallen trees and fell another wide swathe of healthy trees on either side, in an effort to stop bark-beetle infestation. The environmentalists favour leaving the trees where they are, because a rotting spruce acts as a seedbed for dozens of new trees. The beetles are seen as just one of the colourful diversity of bugs and lichen and fungi in the forests, rather than a sworn enemy. In the Low Tatras, and parts of the High Tatras, the traditional response was followed, and the fallen trees were extracted to feed the timber and cellulose industries, leaving whole valleys bare.

'It is not possible to change this way of thinking,' says Matúš. The loggers have been in the business for too long, and there is a lot of patronage. Whole families are involved, dividing the profits up between them. Instead, he and others have been working to change the laws – and the government, when possible – in an effort to protect more trees.

CHAPTER 5

THE HIGH TATRAS

In politics everybody is an enemy. It's a zero sum game. If you attract
votes, I lose votes. And that's not a good environment.

Erik Baláž, Slovak environmentalist

Beyond the Low Tatras, the High Tatras spread like the eyebrows of
Europe. Martin Petráš meets us in the car park in Huty. We – my
Scottish friends Lorna and Dino and myself – are weary after the long
drive from Budapest. It's mid-October, but a lingering Indian summer
offers little threat of rain as we drive up through western Slovakia,
watching autumn quietly paint the green hillsides with streaks of red
and gold. The highways are straight, but reduced to a single lane near
Zvolen, where workmen pour perfect rivers of glistening black tarmac.
Elections are due and the government is sparing no effort to prove that
it has paved the way to prosperity over the past four years.

After Liptovský Mikuláš the road winds up into the hills, with the
huge Liptovská Mara reservoir on the Váh river, invisible in the darkness, over to our right. Thirteen villages were flooded to make way for
the lake in 1974. Before that, archaeologists were allowed to search for
traces of the Celtic tribes that lived here between the fourth and the first
centuries BC. On the hill at Havránok they discovered a major shrine.

The Celts are long gone, and I mourn the view of the mountains rising up all around us, hidden in the folds of an autumn evening. Our road leads over the mountains into Poland, but we take a sharp left to Huty instead. The temperature gauge slips to 5 degrees.

Martin is waiting for us in a Suzuki Samurai, a stubby little off-road vehicle with no registration plates, but a powerful engine and a strong voice. The car park is just a line of vehicles asleep in the mud beyond the end of the village. Dino and I perch on rucksacks and a wooden milking stool in the back, Lorna sits in the front, and we're off, lurching up a steep, almost vertical track into the mountains. The headlights pick out rocks, roots, beech and spruce trees like a stage set, drawn back slightly to let us pass, then closing back to enfold us in the woods. The Samurai cuts a jagged path between the trees.

The bears here, Martin explains as we drive, are milder than Romanian bears, keep themselves to themselves and rarely cause trouble. All the same, we're better off in the Samurai than climbing these tracks on foot after dark.

Down a steep slope into a valley with a fast-flowing river. It's a strange feeling to be in a car driving straight into a river. With childish delight we forge through the gushing waters up the bank on the far side. This is the Kvačianska creek, enlarged by the waters of other streams – the Borovianka and the Hutianka.

A large log house appears in the headlights, lamps glowing in the windows. We've arrived at the mills of Oblazy, in the Choč range in north-central Slovakia – not the highest mountains, but with a special place among the most beautiful.

It's a cold, starry night, the stars only visible in a narrow strip over our heads, while the waters roar beneath our feet. Martin leads us along a wooden walkway, up a step, then into a fairytale scene. A large wooden room, beams across the ceiling, a traditional stove in the corner with a huge enamel pot of water on the boil, a table groaning under the weight of food and wine, and three young laughing women – Jana, Jana and Elena – who welcome us like long-lost friends. The women are locals, volunteers on a rota, looking after the mill for this particular week.

The room is a capsule of heat in the chill forest. We start with plum brandy, then move onto apple. The glasses are tiny, but constantly refilled. Then space is made on the table for bottles of Urpiner, one of the last Slovak beers not swallowed up by big Western brewers, and Zlatý Bažant, owned by Heineken, with the year '73' on the label – 'to give it a retro feel,' Martin says, a little dismissively. There are smoked and curly Slovak cheeses, made from sheep and goat and cow's milk, hard cheeses we brought from Hungary, small red apples from the trees near the mill which the goats normally eat, and fragrant black Izabella grapes from my own garden. At some point, a small bottle of goat's milk is introduced. I sip it slowly, as an antidote to all the alcohol.

Martin is known as 'Pedro' by his friends, after a popular bubble-gum from the 1970s, which sounded like his family name, 'Petráš'. The packaging depicted a boy in a sombrero hat who looked rather like him. The gum cost exactly one old Czechoslovak crown, and the recipe was copied from North America. According to legend, Canadian hockey players on tour in the country discovered it and took vast quantities home. The Czech gum stretched all the way back across the Atlantic to the motherland of all bubble-gums.

Martin and his family grew up in Bratislava, but owned a house in a nearby village and spent long summers there. The Kvačianska and Prosiecka valleys are renowned for their stark, natural beauty. A century earlier, an article in the first edition of the periodical *Beauties of Slovakia* (Krásy Slovenska) stamped this region on the Slovak imagination. On a busy summer's day, a thousand visitors still find their way here.

Some kind of mill has been here for ages. In the nineteenth century, a sawmill was built into the room beyond the flour-milling room, inscribed with the name 'G. Topham, Vienna, 1873'. British visitors told Martin that 'Topham' is a British name, typical of the district around Leeds. The Topham factory was set up in Vienna by entrepreneurs from Leeds.

The mills were in use until the Second World War, tapping the power of the river to grind grain in one section and saw logs into planks in another. The Nazi occupiers of Czechoslovakia rarely dared enter these hills, abandoning them to a motley collection of displaced people and resistance fighters. When Martin's daughter Anna was six, her baby-sitter noticed her playing with a rusty object she had dug up from the mud. On closer inspection it turned out to be an unexploded hand grenade. She was deprived of her treasure and the army called in to blow it up safely.

After the 1948 communist takeover in Czechoslovakia, the mill owners were classified as 'class enemies' and the authorities put a seal on the milling equipment to stop anyone using it. Some locals defied them, but by the early 1950s a jagged red star had been stuck, metaphorically speaking, in the spokes.

Disused and unrepaired, the log building began to fall apart. Rain got in through the roof. The wood in the ceiling rotted, then fell in. Rock slides engulfed the blacksmith's workshop. I imagine the body of a man-made animal, crumbling into the mud, its ribcage still visible.

The Kvačianska gorge, eroded by rain and snow and re-baked by the sun, was also crumbling. Grain from Slovakia was once carried on pack animals on tracks through the mountains to Poland from here. In the interwar years, work began on a road to replace the old track, but was finally abandoned, the efforts defeated by the rock. The trees, picked off one by one for the sawmill, took their revenge. 'I remember this valley like a jungle,' says Martin, recalling his first memories.

When the mills fell into disuse their heavy machinery was not cannibalised, as happened to other mills and factories elsewhere in eastern Europe.

Round the table, our conversation wanders. Tales of the communist past are woven into the present. Martin mentions Hanzelka and Zikmund, a legendary duo of Czechoslovak adventurers after the Second World War, who explored the world in a silver Tatra 87 car, donated by the Tatra company. After the war, few Czechs or

Slovaks had the means or possibility to travel. Undaunted, while still at university the boys drew up plans to explore the five continents, and began learning the languages they felt they might need on the way.

The tales they sent home from their first, three-year journey through Africa and Latin America turned them into celebrities. In the dark and painful post-war years, they seemed to be travelling on behalf of everyone else. While they were away, the 1948 coup brought the Communist Party to power. But on their return, the party decided to embrace them as working-class heroes. So they set out again, this time in a Tatra van, to explore Asia and the Pacific.

Hanzelka died in 2003, while Zikmund passed away in December 2021, aged 102. A museum dedicated to their journeys was created in the Czech city of Zlín. An asteroid belt was named after them: one of those shining down on our mill.

It's getting late, and we're suddenly axed by weariness. Dino and Lorna stay in the snug room, while the three girls, Martin and I climb up to the hay loft in the next building. I stretch out my sleeping bag and get in, fully clothed. The temperature is close to freezing. I put my head inside and breathe heavily. In no time, I've heated up the space and fall into a blissful sleep, lulled by the roar of the river.

In the morning, I'm aware of my nose first, an iceberg in an otherwise warm sea. The others are asleep. I find my way gingerly down the ladder and look around. There's a fine film of frost on the fallen leaves and on the wooden footbridge over the river, so cold it burns the soles of my bare feet. The gorge is like a canyon, the Kvačianska stream narrowing down towards the mill, and I understand the constant roar of the waters. A mini waterfall, over which the waters tumble, has been built into the riverbed. Above and below the step, the Kvačianska is quiet and stony.

On the green shore stands a tall willow, its leaves still green, and behind it an ancient linden, shedding broad yellow leaves: the guardians of the mill. A cockerel struts to and fro beneath them, inspecting the sunlight, its cock-a-doodle-do-ing echoing back off the canyon walls. It is followed by several well-mannered hens.

The mill is composed of a selection of log buildings, topped with a tall, steeply sloping roof made of wooden shingles. Beyond the mill, the canyon narrows sharply between tall cliffs.

Soon Martin and the girls are up, busy with their morning tasks. First, to divert a section of the river into a channel, which flows directly to the giant waterwheel. The waters gush forward, filling it paddle by paddle, 1.6 tonnes of water. But nothing happens. How can so much water not turn the wheel? A creak grows into a groan, then a judder, then a grunt, and the whole mossy contraption shudders into life and turns, ten revolutions a minute. It seems unhurried and, once going, unstoppable. Inside the mill, the girls throw a switch and the lights come on – powered by the watermill.

It charges a car battery, hidden away among the logs. Everything here is created from water. I go to help milk the goats, tripitty-trap across the wooden bridge where my morning footprints still form strange, humanoid islands in the frost. There are two mother goats and several snow-white kids. Their living quarters, a large wooden shed awash with hay, is pungent and homely. The kids are allowed out for their mothers to be milked.

Jana attaches a thick red cord from the nanny goat's neck to the wooden door, in case she starts moving around too much; then she manoeuvres a white tin bowl between her hind legs. She shows me how to do it, squeezing the udders rhythmically, squirting the milk into the bowl. Then it's my turn. It's many years since I last milked a goat, though I still remember her name – Amalthea, a beautiful, temperamental creature on my friends' farm in Greece. I was never very good at it, too hesitant, as though a male of any species shouldn't be doing this to a female. But the girls are waiting – and so, it seems, is the goat. So I kneel down in the hay and take hold. Her udder is warm and leathery in the cold morning, good in the hand, but not as full as I had hoped. I give a nervous squeeze and nothing happens. You have to pull down and squeeze at the same time. This time a small squirt, then another. I'm getting the hang of it, but I'm afraid of hurting her. The girls take over, and I retreat gratefully.

When the nanny goats are milked, they troop over the bridge with their kids to the mill side of the river – another scene from a fairytale 'Three Billy Goats Gruff'.

The restoration of the mill at Kvačianska was one of the great successes of the 'back to the land' movement, described by Mikuláš Huba. 'The 1970s and '80s were a time of powerlessness and stagnation in this country. The party had absolute power, and people felt they could not influence their own lives. Young people began to restore old buildings, and show that something could be done. That impressed the local people.'

The communist authorities were nonplussed by the energy of the long-haired men and women who got involved. The first group of volunteers came here in the early 1980s to restore the timber-framed buildings and the roof.

Lego was Martin's great passion as a child, and he made models of the mill. Later he became a civil engineer, specialising in bridges. He wrote his dissertation at Bratislava University on the wooden footbridge across the Kvačianska valley – 'the teachers were surprised, everyone else chose big steel suspension bridges . . .'

At the age of fourteen, he wrote to the club set up by Peter Kresánek to ask to take part in the restoration. The huge old wheel had to be extracted from the mud. There was also a lot of simple digging to do. Landslides had buried parts of the buildings under rubble.

Under communism, 'state commissions to protect the environment' existed in each country, but their hands were tied by the unrestrained industrialism of the party. While dissident groups like Charter 77 in Czechoslovakia confronted the regimes over human rights, the environmental movement grew, quietly restoring a building here and there, attracting less attention.

We walk in the late afternoon along the track to the village. The sunlight is intense, flooding straight down the gorge like water, as if to make up for all the times when it shone on the rest of the world, but not here.

⊙ ⊙ ⊙

My initiation into the High Tatra mountains takes place the following May beside the Belá river with Erik Baláž, as the jagged peaks play hide-and-seek in the clouds.

Thin and fresh-faced in a green jacket and pale trousers, Erik looks younger than his forty-three years. He's one of the best known of a generation of conservationists in Slovakia. He has recently quit politics and is now back in the wilderness, doing what he loves. 'There is so much water in the mountains,' he says, as we cross a small footbridge into a wetland forest towards the Belá river. 'In the Tatra mountains, you have up to 2 metres of rain and snowfall a year, and only 300 milli-metres evaporate.' Down in the Danube valley, he explains, there are just 300mm to 600mm of precipitation a year, and 1,000mm of evapo-ration. The land is drying out. 'This is the balance between mountains and lowlands. Life in the lowlands is only possible thanks to the mountains, where there is much more rain.'

There are several channels with water running through the woods, and many pools where the water lies still, pond-like among the roots of the trees. The Belá river is constantly changing direction. Almost alone among the rivers of Slovakia, there are no dams at all on its 22km, before it flows into the Váh at Liptovský Hrádok. The woodlands on its shore store the water, and it never dries out, even in the driest of summers.

> I grew up in Liptovská Kokava, and spent all my summers by this river. We drank from it and swam in it, and we still do. Each year at the beginning of summer I walked with my father through the river to find the best places to swim, because those places are quite rare. That's when I understood that the wild is important.

In summer, the warmest temperature of the Belá is 12 degrees Celsius, while in Liptovská Kokava it's 16 degrees. In summer storms, hundreds of thousands of cubic metres of water pour down from the mountains and find a place in this forest, rather than flowing away downstream. When the water level in the river falls again, water stored underground returns to the river.

On his phone, Erik shows me a graph of floods in the Belá in the twentieth century: a big one in 1928, following huge clearcuts of forest in the Tatra mountains; and another in 1941, after more clearcuts. 'These graphs show that a landscape which was destroyed once and is then well managed, can recover and reduce flooding – thanks to better water management and conservation.'

By now we have reached the river shore. It's mostly stony, but with just enough depth on the far side for a dozen rafters to launch their dinghies. It's some kind of rescue practice, with volunteers swirling out into the current in lifejackets, only to be hauled out by their colleagues. The water-sports people are important allies, Erik explains, in his battles against new hydro plants.

He makes documentary films with his friend Karol Kaliský about the wildlife of the Tatras, spending weeks camping in the mountains to get the best shots of animals, in all four seasons of the year. The films are spectacular – bear cubs playing with their mothers, and climbing trees to gather pine nuts.

Erik was one of the founders in 2017 of a new environmental group, We Are Forest (My Sme Lesy). At that time Robert Fico was prime minister, and his SMER party had been in power for six years, with policies which strongly favoured the logging industry. In March 2018, Fico was forced to resign, following the murder of a young investigative journalist Ján Kuciak and his fiancée. Fico's SMER party colleague Peter Pellegrini took over as prime minister until the 2020 elections.

In the meantime, Erik Baláž entered politics, and joined a new party, Together – Civic Democracy (SPOLU), becoming one of its vice-presidents. 'It took us twenty years to save two valleys from logging in the High Tatras. I thought we could do it faster as politicians,' Erik explains.

Zuzana Čaputová, a lawyer and environmental activist involved in saving the Slatina river, was elected president of Slovakia in June 2019. From March 2020, human rights and green activist Ján Budaj served as environment minister. The green movement in Slovakia enjoyed more influence under his leadership and with the president's quiet support than ever before.

In the 2020 election, in alliance with Progressive Slovakia (PS), Erik's SPOLU party failed by just 900 votes to make it into parliament. He quit in 2021, after three years in mainstream politics. In September 2023, Robert Fico returned to power, and new conflicts over land use.

Erik laughs about his years in politics now, as a strangely disempowering experience:

> It was very good to understand how it works. Maybe we could have changed a lot. But once you're in politics, you cannot do anything else. You have no access to the media, for example. If you want to influence something in conservation, you have to have a voice . . . If you have a party with 6 or 7 per cent, and you are in government, you are the last one in the line. They can always say, 'I will not give you money. I don't want to change legislation for you or give you EU funds.' When I didn't see real possibilities to succeed at that really high level, I jumped.

During the pre-2020 governments under Prime Minister Robert Fico, around half the land in the national parks belonged to the state, and the management tried to extract maximum profit from them. After 2022, the government used funds from the European Union's post-Covid Recovery Plan to reform the management of state and private forests. Paying off individual owners year after year not to harvest the trees struck policy makers as an inefficient system. The new concept was to privatise certain areas, and place them under strict protection.

When Erik and his friends took Slovakia to the European Court for failing to protect its forests – and won – the Slovak state evaded the fine by promising to introduce zoning plans in the national parks, on a scale of 1 to 5: Level 5 is the most protected, with no logging, construction or hunting allowed.

The aim was to gradually increase the area of forests under strict protection from 2 to 10 per cent of Slovakia. Budaj's ministry also drew up plans to revitalise the waterways, make houses more energy-efficient, and reduce rural emigration by creating more jobs in eco-tourism.

'Many of the regions where we have national parks are poor, the people do not have enough work,' says Erik. 'If you try to stop them making money out of forestry, they will be angry.' Reforms should be accompanied by investment in education and tourism:

> Every national park should build its own school for kids. These will be crowded all year around, and create employment for the villagers. We have nine national parks, so we need nine schools each costing €10 million. So we need €90 million from somewhere. If we don't have that, we cannot even start to talk to local mayors.

In his vision, the national park rangers would become tour guides, working with tourists and groups of children, bringing in money.

The national parks have 'core zones' enjoying maximum protection, currently 10 to 50 per cent of their whole area. The state foresters opposed zoning, because it encroaches on their freedom to move at will through the parks, cutting and hunting as they see fit.

In practice, says Erik, 'there are only two zones – Level 5 and the rest'. He would like 75 per cent of the national parks under full protection:

> I try to explain to the people from the state Nature Conservancy Ministry that we have to protect compact areas like entire valleys. If you have no logging in the valley, you have no mud in the water, because there is no soil erosion. There are many scientific arguments for this approach.

The Belá flows crystal clear, and we crouch to drink from it, the water cupped in our palms. Erik is not worried that if more people come, the pressure will be too great on the wild places:

I think this is usually a false dilemma, because people don't like to meet other people in nature. They like to feel that they are alone there, in the wild . . . It has to be well managed, with no roads and no cars. Before it was a national park, people used to go everywhere. When you declare an area a park, it can be better protected . . . You don't need to manage forests anymore. Imagine a path through the Tichá valley, 17km long. And there are secondary valleys without paths which are 6–8km long. Nobody will go to those secondary valleys. And the bears will always have a peaceful environment, with fallen trees, undisturbed. Then you can ask a guide, please take me there. You pay some money. The guide knows how to enter the valley, where to cross a river and where you can observe the wild animals.

One September, Erik was here alone, planning to spend the night at the foot of one of the tall, ancient stone pines below Kriváň. He kept bumping into bears, which – to his surprise – did not flee far. When he tried to sleep, he was woken by two bear cubs chasing round and round him for a pine cone. Then one of them shot up the tree and started eating pine nuts. Erik didn't get any sleep that night.

Kriváň mountain was long believed to be the highest in the Tatras. The crooked top created the optical illusion that it was taller, sailing out from the clouds like the tip of a pirate's sail. There's a local legend that links the peak to the region of the Lipto ethnic group. The devil was flying over, dragging with him a new batch of sinners bound for hell, when he caught his foot on Kriváň, bent the topmost rock, and dropped his precious cargo. Saved from the fires of hell, they escaped and have been thriving ever since in the foothills, counting their lucky stars at their miraculous escape.

We come to a 'blind arm' in the river: the water is constantly going underground, disappearing and reappearing, purifying itself as it is filtered through sand and fine gravel.

Since the first hydropower station was built on the Hnilec stream in 1912, most rivers – large and small – have been dammed, some of them many times. 'Many people used to travel by boat on the Váh,' recalls Erik. 'From spring to autumn it was full of people, fly-fishing, on guided tours, looking for accommodation. The rivers brimmed with fish. Then they built three dams, and everything collapsed . . . For the sake of very little electricity, they destroyed many rivers in Slovakia.'

On the way back to the car, through the flooded forest, we talk about belonging: 'Here in Slovakia, we have the same language, but we lost our history. We were part of the Austro-Hungarian monarchy for hundreds of years, but we don't talk about that. We just decided that was not our state.'

Many towns and villages of the valley boast the name Liptov: Liptovský Mikuláš, Liptovská Teplá and Liptovský Hrádok. Erik doesn't feel any strong Slovak or Lipto identity:

It's difficult. I'm connected to nature, not to people. I have many enemies here. Because I'm a conservationist, half the Lipto population want to kill me. Many are against me, while some are on my side. But I don't feel part of this community.

I have a community but they are not all located in this specific place. I have friends around Slovakia. We meet quite often, we come here and sleep in the forest and talk, or go to Romania together. This is my community, my family. Often if you live in a small village or town, it's hard to find enough people who are interested in the same things as you are.

When I first came to the Tichá valley, I was eighteen years old. And I told them, 'okay, stop. This is a nature reserve, a national park. The bark beetle is okay. You don't need to plant trees, that's stupid. They will grow by themselves.' I tried to explain that we need deadwood, because there are woodpeckers and capercaillies. They need the structure of the forest, these lichens, and the soil and water. But when they heard all this, they just told me to get out.

They're afraid of losing something – their position, their land, their hunting rights, because of conservation. So they will never accept it. But I cannot say 'okay, just do it'. That is not possible.

In my experience, if I can take somebody who is angry with me to the forest for a one-day trip, we can speak about all the topics from different sides. He will not necessarily change sides, but at least he will no longer be my enemy. But if they recognise me from social media or from television, it's not like that.

I do a lot of walking with important people, sometimes from the region, but other times politicians, journalists, scientists, or business people. We go to the forest, and I try to explain why it is important. And that's why we have a much bigger coalition now, backing conservation. But that coalition is not from the village, it's from the whole of Slovakia.

Ľudovít Štúr and his friends made it to the top of Kriváň on one of their excursions in August 1841 – the first recorded ascent to include women. There's a metal plaque, nailed to a rock at the summit, placed there by the Matica Slovenská organisation, next to a blue waymark.

In 1839, Ludwig Greiner was the first to prove that Gerlachovský was the highest peak in the High Tatras and the whole Carpathian range. He climbed them all and used triangulation points, instead of an altimeter. According to the latest measurements, Kriváň is 2,495 metres high, Lomnický 2,634 metres, and Gerlachovský 2,655 metres – the narrow winner, if mountains compete with each other.

The wetland trees in the Belá valley also get a little help from the beavers. *Bobor* in Slovak, the beaver had all but vanished from Slovakia by 1910. In 1977, isolated individuals were spotted, and by 2012 there were stable populations. The beavers in the Carpathians are a mixture of native beavers (which survived centuries of hunting and culling) and the product of the rewilding programmes that have been pursued in Austria and Bavaria since the 1960s.

'Nobody can tell where the river will go next,' says Erik. 'It all depends on where the trees are when it floods.' Fallen trees and branches clog up one artery, forcing the water to curve around them – unless the trees are dislodged by the force of the current, and the water goes back to the previous bed. The same happens, on a micro scale, with the dams the beavers make: the water trickles through them and they just slow it down. The beavers made the whole landscape of Europe. When politicians in Brussels speak about the 'architecture of Europe', perhaps they should remember the original beaver architects.

The Belá flows, in quiet times like this, at 7 cubic metres a second.

When there is a low water level in summer, the river moves only clay. So you have one level of sand, then the level of sand and clay. And after millions of years, it goes down to the ocean. There is big pressure as it pushes all these sediments. And it makes this kind of stone. You can see the layers on the opposite shore. This was a delta river, tens of millions of years ago.

Lingering beside the small bridge over the stream, I ask Erik if he believes in a Designer – in God.

'No, it's just evolution,' he says, disappointingly.

The forests can store a lot of water, but how was that created? Because trees need water. So they try to keep water for themselves. They produce chemicals, which creates a good soil structure. And the same management design is used by fungi, because they produce something which stores water, so they have water for their own use. So in a sense, different parts of the 'society' of the forest work on an egotistical principle, for their own survival.

'But,' I suggest, 'it seems like they're thinking about it. That the different parts of the whole have an innate intelligence.'

Yes, in the sense that nature can solve problems. But the power of evolution is very strong. In many small changes you can find the best solutions. So you have artificial intelligence to create a Boeing aircraft in a computer, you have small changes in all the possible directions, and evolution can find the best solution for that time.

'But if there's no spirit, no soul,' I prompt, 'then we can be replaced by robots, because if robots are so good, they can self-correct. Are we more than that or not?'

I think we are more than that. We humans are ecosystems. We are not much more than a bear or any another organism. You cannot make a mobile phone from a stone, can you? But society is able to make it, because it has a higher level of organisation.

Think of your gut. You have thousands of species of bacteria in your digestive system, all doing different things, adapting to whatever you eat. No single one of those bacteria is very intelligent, but together they make something very sophisticated. Evolution is all about that. So you have small things happening, billions of things happen every second in this forest, solving problems. And as a result, you have a complex ecosystem. Thanks to billions of cells in your brain you're able to think. And it is not managed by one central cell at the top, directing you to do this or that. Nobody is intelligent enough to create your body. DNA is not so intelligent.

'Don't you feel that the Belá river has a soul or spirit of its own? A personality?'

Yes. I have this feeling of spirit. But to me, it is just this secret and complexity and the long period it has been here. I feel respect for all these things which have been here for millions of years. But I don't feel that there is a single spirit, like God, or the personality of the river as a single river, I think it is like an ecosystem which is constantly changing. It is not the same river I saw when I was young.

Erik's strongest spiritual experiences were in his late teens:

I spent a lot of time in the Tichá valley, mostly alone. I just sat there on the top of the hill. And I could see all around me. I watched bears and deer, and the birds flying and the weather changing. I didn't care about myself. After three days, you forget that there are people down in the valley. You are just alone. And you have no purpose. You don't want to make a film, take a photograph or write a book. There is no reason to be there, but you are there. You are not important. You are part of the ecosystem. You are the Tichá valley. You are the universe. I didn't feel separate. I was not afraid. I was very strong. I was able to imagine everything possible. Usually people think in small boxes. But at those times, I was like a really big brain, thinking about it all.

◎ ◎ ◎

It's snowing when I reach the parking lot at Podbanské. The snow blows in vertical lines through the spruce trees and the wind almost wrenches the car door from my hand. It's mid-September, and Martin Mikoláš and I plan to hike the Kôprová (Dill) valley, parallel to the Tichá valley, up towards the Polish border, beneath the pointed fringes of the mountains.

Martin is a dendrologist, a scientist who studies the age of trees by counting the rings in the trunk, one for each year of its life. He measures the age of living trees by drilling a small tube through the centre and extracting the wood. But when the wood is rotten at the core, as is the case with most very old trees, the process is harder. One option is to take a sample from a large branch. Each ring is carefully studied for information about the weather conditions in a certain year or span of years, and as many trees as possible in a forest are studied. Subtle changes in climate, in the activity of beetles and other creatures which attack the trees can all be found. The trees are the historians of forest and mountains. One just has to learn their language.

We climb up through the lower slopes, still scarred by the massive windthrow of 2004. Martin leads me off the track to examine a perfect example of the way forests regenerate by themselves, if left in peace. The long, prone body of a spruce which fell at least forty years ago is now barely distinguishable from the rest of the thick undergrowth. A long line of young spruce trees sprout from its length.

The forestry company argued that it would take centuries for the forest to regenerate after the storm. In areas where they cleared the fallen trees, that may be true. But where the fallen trees were undisturbed, there is already huge regeneration.

We climb the steep path towards the summit of Kriváň. The High Tatras are the knuckles of a clenched fist of sharp, jagged consonants – Kriváň, the spectacular waterfall of Vodopád Skok, Gerlachovský štít, Mengusovská, Batizovská, all scraped by extinct glaciers – all those 'k's, 'v's, 'g's and 'z's spat out from some ice-age mouth.

Even on a Wednesday morning in September, there are many other hikers. A couple carrying their small children – an eighteen-month-old on his father's back, his three-month-old sister on her mother's front – pass us at high speed. 'Is this her first ascent of Kriváň?' I ask. 'Yes, indeed,' the mother replies, hardly pausing for breath. 'Got to start them early!' 'And how is she enjoying it?' I call after their receding figures. 'Slept through it all so far!'

By now we have reached the undisturbed part of the mountainside. After years of direct action and court battles, the upper Tichá and Kôprová valleys have won strict protection. Rowan trees, *Sorbus aucuparia*, their long delicate leaves reminiscent of acacias, grow thickly around us, among the stumps of wind-toppled spruce. The rowans are a pioneer species, much disliked by the forestry companies as obstacles to their clearance plans, because they spring up rapidly in areas affected by storm damage. But in their shade, young spruce trees find perfect conditions to grow, elbowing the rowans aside as they do. The rowans die at about forty or fifty years, but the spruce keep growing for decades or centuries more. The complex interaction of spruce, pine, rowan, bark beetle and all the intermingling species on the forest floor ensures the growth of a healthy, mixed-age forest, more resilient to future storms.

Far below us, we catch sight of a curve in the Belá river. The Kriváň peak appears momentarily from the clouds to our right, as if to check our progress – ants on its shoulder. The morning sun highlights a scattering of snow on the peak. Then it buries its head back in cloud. There is snow among the bilberry bushes, too, like icing sugar on the dusky blue berries.

The track is deep and stony, and my hiking poles are almost useless. Long lines of cloud scud over the valley, grey and white, with little threat of serious snow. The sun comes out sporadically. Martin is a mountain runner, and nimbly leads the way forward.

The Tichá and Kôprová valleys collect the streams which form the Belá river far below. Martin stoops to identify bear scat, russet brown and fresh, full of the tell-tale traces of pine cones. Round the next bend in the track we come to a 'bear tree'. The bark has been rubbed away to the core by a succession of bears, and their hair is clearly visible, glued to the resin. This is not just a casual scratch on a long lope through the mountains, Martin explains, but is the subtle way in which the bears leave messages for each other. Bears who want to make their presence known urinate at the foot of the tree. They then dunk the fur on the top of their heads in the urine and rub their heads on the tree, as far up as they can reach. New bears, arriving at the tree, sniff it carefully. From the scent, a female who wants to mate can tell much about the bears who have passed this way. A mother bear with cubs to protect from aggressive males leaves no trace, and simply hurries away with her cubs. In the same way, wolves and bears can tell from the droppings of deer and other potential prey how young and fit, or old or ill, the deer is, and therefore whether it is worth pursuing.

The first stone pines from which pine nuts are taken, are magnificent trees. Erik and Karol have filmed bears scampering up them to harvest the precious, nut-bearing cones. All pine cones contain some nutrition (as squirrels know), but only the stone pines in Europe provide useful-sized nuts. The pine nuts of Italy have been ravaged by the Western conifer seed bug, *Leptoglossus occidentalis*, imported accidentally from North America in a load of timber in the 1990s. The stone pines of the High Tatras, in Slovakia and Poland, have mercifully been spared so far.

We eat our sandwiches in a lynx's den. The brown needles are flattened beneath an ancient stone pine – a perfect lookout point for this shiest of cats into the valley below. Lynx faeces – none too fresh – are piled to one side of the den. Fallen trees to the edge offer more cover, and there are gaping holes beneath the trunks where lynx cubs might hide.

We scramble down a steep slope. There's no track, and progress is slow through the thick undergrowth. It would be easy to twist an ankle or tumble down the mountainside. After half an hour of clambering, downwards and sideways, we come to a clump of three enormous trees, old giants, leaning on one another. Closer up, their stone-like wood is curved and cragged, their foliage voluptuous, the nuts in the cones at their feet delicious. The bears got here first – many of the cones are broken open. But there are so many – plenty left for us, too. Martin reckons the oldest trees could be a thousand years old – the oldest in the Carpathians.

⊙ ⊙ ⊙

In May 2024, I return to the Tichá valley with Martin, Erik and a group of scientists and researchers, to explore the idea of rewilding. Until then, I had understood this to mean the reintroduction of bison, beavers and wolves to places where they had not been seen for decades, or centuries. On this trip, I came to understand it as something nature does herself – if we simply stop interfering.

On an early-morning walk in the forest near Podbanské, we identify chaffinches, chiff-chaffs and three kinds of thrush. I wash my face in the ice-cold water of the stream and risk a delicious mouthful.

After breakfast, we set out through the Tichá valley, roughly parallel to my hike with Martin a year earlier. Taken together, and spilling over onto the Polish side of the mountains, the area comprises some 500 square kilometres of wilderness – one of the largest such expanses in Europe, apart from some places in the far north and in the Alps.

Still on the tarmac, we spot the first bear scat – as large as a small cowpat, with a patch of fur clearly visible in it, probably a chamois, one of the small, wild goats that still roam the Tatras. 'Bears need a big area of wilderness, to find different sources of food all year round,' explains Erik. 'If there are too many people, they feed on cornfields, or rubbish. In this valley, they just do their own job, naturally.'

One of the scientists asks what Erik would like to happen next, from a rewilding perspective. The strictly protected area should be extended, all the way down to the highway, he replies. At the moment, in winter the deer and wild boar go down into the valleys, because of the snow. If hunting was banned in the lower reaches, too, there would be more for the bears and wolves to feed on, and bison could be reintroduced, too. Wolves would kill the weaker animals and the bears could feed on their carcasses, as in Yellowstone Park in the US. The herbivores would change the vegetation, and the area of wilderness could grow.

When he started the campaign to stop logging in the Tichá valley, aged eighteen, the authorities told him 'OK, just the pristine forests.'

'What they didn't understand, is that you can't just have small patches of wilderness.' If they had won, there would be no wilderness in the Tichá valley now, and the waters of the Belá would run constantly with mud from logging upstream.

Erik cites the example of the capercaillie, the astonishing wild hen, also known as the wood grouse, which has been driven to extinction in much of the Carpathians. The birds are large and cannot survive in dense, managed forests, where they crash into trees when they fly. They need relatively open forests where the branches come all the way down to the ground, so that they have places to hide from birds of prey, at the base of trees, under the snow. Scientific studies show that capercaillies need an optimum area of 100 hectares per bird; so even an area of 2,000 hectares of suitable forest is not enough, as the twenty birds living there will die out from the effects of in-breeding or predators. On a positive note, the capercaillie has already helped save and expand the Slovak wilderness: several thousand hectares in the Great Fatra National Park have already been set aside, in response to a ruling of the European Court – all thanks to the capercaillie.

We climb higher, off the path now. Beside a fallen spruce, rotting slowly into the undergrowth, Erik explains the mystery of carbon:

> There is as much carbon in the forests of the planet as in the atmosphere. But if you log trees, most of the wood is burnt. When trees die and are left to rot slowly in the forest, some carbon is released, but most stays, especially in cold and wet conditions. Pristine forests retain much more carbon than commercial forests. If you make a clearcut, carbon is released when the timber is burnt, but from the remaining soil as well, because it heats up and dries out.

Dead wood also acts like a sponge, first absorbing the water, then releasing it to sustain the living trees, in dry weather.

> We believe that this is one of the most important topics from the perspective of rewilding. If you leave the forests, which were managed before, they will store more and more carbon in the future. And that means less carbon in the atmosphere.

Natural calamities always existed here, adds Miroslav Svoboda, professor of forest ecology at Charles University. Windthrows, avalanches and droughts. This is confirmed by the village chronicles, dating back over hundreds of years, and by the work of dendrologists and paleo-ecologists, who study the pollen in the peat bogs around the high mountain lakes.

As we climb, we leave the beech trees behind and move into the realm of pine, fir and larch. Then the stone pines – not as huge as those I saw with Martin, but amazing nonetheless. We spot a male three-toed woodpecker, *Picoides tridactylus*, with a yellow cap. Martin mentions that the wood of the stone pines was traditionally favoured by couples for a bed, as it did not creak when they made love.

Then we walk along a ridge, with Kriváň peak – still covered with patches of snow – to our right and with the peaks of Poland straight ahead, above the lakes of snow-melt at the top of the valley. Here we come across the first scat with wheat ears in it – proof that the bear had eaten food left out by foresters for wild animals. Earlier on the walk, we found the fur of a chamois – a mountain goat – in a scat. We have counted around two dozen bear scats between the huts at Podbanske and here. To the expert eye, the others mostly indicate a diet of grass or pine tips: 'The bear is looking for nourishment. A tree sucks up the sugars through its roots, and they rise to the top of the trees, so the bear climbs to the top to chew the branches.'

The sun comes out, and we pause to eat our sandwiches. Someone passes round slices of 'copper cake', a spongy French delicacy. Then Martin calls out from the ridge ahead. We take turns with his binoculars to watch a mother bear amble up the opposite slope with three small, darker cubs. They're grazing on the flowers of the bilberry, taking care to keep away from any large, alpha male who might kill the cubs to get them out of the way, so that he could mate with the female. No doubt, the mother had sniffed that bear tree we saw earlier and hurried her cubs past it, leaving no trace.

CHAPTER 6

AND TIME STOPPED UNRAVELLING

And when people cease to believe that there is good and evil,
Only beauty will call to them and save them . . .
　　　　　Czesław Miłosz (1911–2004), Polish American writer

There are two Galicias in the world: one in Spain, the other in Poland. The Polish Galicia is probably named after the medieval city of Halych, now in Ukraine. The Kingdom of Galicia and Lodomeria was on the southwestern fringe of the Polish–Lithuanian Commonwealth. Many have claimed Galicia, but in Poland the *Gorale* (which means simply the 'highlanders'), in their mountain fastnesses on the northern slopes of the High Tatras, have pride of place.

The bus from Budapest to Kraków leaves at one o'clock in the afternoon. January again, snow promised in the mountains, beyond the heavy rain of the city. Our first real snow this winter. The bus cuts north, across the Danube plain of southern Slovakia to Zvolen, where the first flakes turn the roadside white, then Banská Bystrica. By now it is dark and each bus station is uglier, damper, sadder than the last. The bus began its journey in Zagreb, the drivers are Croat, and they radiate a strange contempt for their passengers. One talks all the way, hour after hour. I barely understand a word, and his voice sounds like a disgruntled river flowing nearby.

The highpoint of the journey is the mountain pass at Donovaly in the Low Tatras. The snow forms mountains on the pavements and the road turns into a hazardous trickle of vehicles, where diligent snowploughs are losing the struggle. An American girl gets off alone, lugging her suitcase into the snow-spangled night in her woolly hat, with a parting 'Ciao!'.

After Donovaly, the snow thins out. We drive down into the valley of the Váh at Ružomberok, then up again into the western Tatras, towards Dolný Kubín and the Polish border at Jabłonka. We creep forward behind slow-motion trucks and cars in snow-light. We reach Kraków only half an hour later than scheduled.

First impressions are not promising: a sea of bright lights after the magnificence of the mountains, and a jungle of concrete underpasses and overpasses beside another subterranean bus station. We get lost immediately in the streets, in search of a tram stop. Eventually, the number 14 takes us to the street named after Stefan Batory, king of Poland and prince of Transylvania in the late sixteenth century, when Kraków was the capital of Poland.

We wake to the hush of snow-covered roofs, and venture across the main square into the old town. The amber jewellery in the cloth market glows gold and auburn, turned darker by the snow outside. We walk past Wawel castle, where Batory is buried, on its limestone hill on the left bank of the Vistula.

In the former Jewish town of Kazimierz, despite the friendly cafés and restaurants, there is a heaviness in the air. In the courtyard of one, there are Roman Vishniac photographs from the 1930s on display, documenting the Jewish world lost in the Holocaust. A school group from Israel huddles in one corner of the square, the girls all in long dresses and black coats – even their woolly hats are black.

The Vistula looks dark and bleak, trapped and regulated, a silent witness in the city. An elderly man shovels melting snow into a wheel-barrow: it could be a scene from a Vishniac photograph. We escape from the melting streets of Kazimierz into the bright warmth and colours of the Ethnographic Museum.

There are timber houses from Podhale, with small windows for people who spent most of their lives outside. The white felt trousers worn by the men, and the colourful printed or embroidered skirts of the women line the walls. The only other visitors are a tall, thoughtful Polish man wearing glasses, his shy wife and their exuberant little daughters, their blonde hair tied in beautiful plaits that almost belong in the museum. The man notices us studying glass cups, for 'cupping' a person's back to relieve pain and stimulate the immune system. 'My grandmother used those,' he tells us. 'I use them today,' my wife replies, to his amazement.

There's a photograph entitled 'Two Women Looking out of a Window' – taken in the 1930s in Wetłyna, an ethnic Boyko village in the Bieszczady mountains. The women are grinning happily, as though they know the photographer well – or maybe he has just said something flattering to make them laugh.

The story of the Boykos is a mysterious and often sad one. They are the descendants of an eastern Slav tribe, the White Croats, who made their home in the Polish and Ukrainian Carpathians. Before the Second World War there were an estimated 400,000 Boykos in the foothills. 'Those Boykos are the most mysterious tribe to be found the length and breadth of the Carpathians. No one else is quite so troublesome,' wrote the Ukrainian novelist Taras Prokhasko.

> The Boykos are a little mute. They are incapable of talking about themselves. They call themselves Verkhovynians ['highlanders' in Ukrainian], Rusyns, Galicians but not Boykos . . . They are so vivid when you are among them, yet become slippery, like their waters, when you try somehow to define them.

Almost all were expelled from their homes in Poland in the forced relocation of 1947, suspected of loyalty to Ukraine. The Orthodox church and all 152 houses in Wetłyna were burnt down. Many ended up in the Soviet Union in the population exchange of 1951. In the

Polish census of 2011, only 258 people identified as Boykos, of whom just fourteen gave it as their only national identity.

Upstairs in the museum, there are painted glass icons and a carved wooden Mary deep in prayer. Beside her is a magnificent 'man of sorrow' from a roadside shrine, dated 1650. His face rests on his right hand, like my mother in her hospital bed, during her last days on earth.

While the Boykos were forced to leave Poland at a moment's notice, millions of Poles were forced out by poverty. Between 1888 and 1939, nearly 5 million people emigrated, mostly bound for North or South America.

The Lemkos are another ethnic group in the northeast Carpathians. Ukrainians regard them as fellow Ukrainians. In the 2011 Polish census, 11,000 gave their nationality as Lemko. In the years before the Second World War, Poland, fearing a resurgence of Ukrainian nationalism, encouraged a separate Lemko or Ruthenian identity. Like the Boykos, they sometimes identify rather as Ruthenes from Ruthenia – the westernmost Ukrainian province of Transcarpathia.

'In April 1946, the population of Wisłoczek was escorted to Ukraine by the Polish army. The buildings were burned down.' In 1939, the village had 770 inhabitants, and boasted a wooden church, a mill, an inn and a library. Its ruins were left to fall apart for more than twenty years, before Pentecostalist Christians moved in in the 1970s.

Among the photographs in this section of the museum is one of tent-dwelling Calderash Gypsies near Kraków. Taken by the artist and ethnographer Walery Eljasz, it shows the crowded mouth of a tall tent, in which a man is hammering the base of a copper bowl into shape. Several women, a child and another man gaze into the camera. In the background, the towers of Kraków are clearly visible. The picture might have been taken on the shore of the Vistula. There's a humble, juniper-root basket at their feet.

The musical section of the museum is less melancholy. It contains a marvellous collection of shepherds' trumpets (*trembita*) and short horns, violins and flutes used by the mountain people to communicate across the lonely valleys, or to entertain one another on the frequent feast days. One photo shows a wedding band in the Podhale region in the mid-1950s. Another a blind man playing a hurdy-gurdy in the Hutsul region in the 1920s, a double flute and a selection of bagpipes. The traditional local fiddle, the *złóbcoki*, has only four strings. There are Bethlehem players, dressed in all manner of strange masks and costumes, including New Year's Day carollers from Zwardoń:

> Carollers rushed into farms making a noise and doing mischief. Such chaos and harmony were then stopped by cracking the whip and ringing the bells which symbolically restored order. Animal monsters – especially bears and goats – emanated a special energy, constantly prancing around.

Some of the masks parody Jews. According to the museum's curator, Magdalena Zych:

> The dehumanisation and mockery visible in the masks is the first step in pushing the limits of consent to violence against those who live in a different way. The conventional manner of presenting the Jews and Roma reveals the destructive power of stereotypes. Is this tradition really innocent?

Or could such masks have a positive role, exercising (or exorcising) imagined animosities, allowing people to let off steam? I'm not sure. There is little understanding left of ritual in the politically correct 2020s.

On the top floor is an exhibition of 300 works of art which convey the spirit of the eminent twentieth-century poet Czesław Miłosz's 1986 collection *Unattainable Earth*:

> So that for a short moment there is no death
> And time does not unreel like a skein of yarn
> Thrown into an abyss.

One exhibit shows all the brightly coloured threads from the weaver's craft:

> To whom do we tell what happened on this earth, for whom do we place everywhere huge mirrors in the hope that they will be filled up and will stay so?

There's a naive painting by Eugeniusz Krawczuk from 1981, of a dockworker wearing the hard hat of the Lenin shipyard in Gdańsk – where the Solidarity trade union began – walking through a rural Polish landscape, holding a tiny Polish flag on the end of a stick. Four ferocious red dragons are visible beneath his docker's boots. In the background, all the trees are leafless.

> We don't know the heavens and hells of the passers-by.

This Miłosz quote is juxtaposed with a woodcut of Jesus, poking a devil with his staff. Nearby is the Holy Trinity, in which God looks unusually father-like, with his arm draped affectionately around the shoulder of a young Jesus, while the two men point at the Holy Ghost, represented by a white dove, perched on their knees. There's also a simple painting, tempera on paper, called 'Mother's Death', by Monika Lampart from 1934, in which a mother is laid out on her bed, with relatives kneeling beside her, and two angels in blue, their wings outstretched arriving to take her soul. A devil-like figure cringes by the door.

The bus to Zakopane leaves Kraków at 08.30 on a Sunday. The snow is melting fast in the city, and we're keen to get back into the mountains.

The broad, two-lane highway starts climbing almost immediately. The foothills are thickly populated with small towns and villages, through Myślenice and Nowy Targ. After two hours, we're stuck in a traffic jam into Zakopane. It's the height of the ski season, and Zakopane is the most popular resort in the country. Skiers zig-zag down a mountainside beside the road. A river called the White Dunajec is distinctly black with the melting snow.

In the first years of the nineteenth century, the Polish geologist Stanisław Staszic began to explore the 'less hospitable northern slopes of the Carpathians, more often than not gloomy, rainy, cold – if strikingly beautiful.' This was a traumatic time for Poland, which had just ceased to exist, partitioned in 1795 between the Austrian, German and Russian empires. 'You enormous cemetery of past centuries,' Staszic addressed the mountains, 'you most lasting monuments for future centuries, your peaks filling the clouds, you will preserve the indestructible name of the Poles.'

The paradox was that the Polish landscape and the word 'Poland' came from *polje*, meaning a field, or flat landscape in proto-Slavic. But the flatlands were easily conquered, and the peasants easily corrupted. From the late nineteenth century and into the twentieth, the Carpathian mountains along the southern border were seen as the strongholds of 'Polishness' – the fringes of the country that would survive the national humiliation. Zakopane was transformed from an insignificant mountain village into the 'Polish Athens'.

We get off the bus at the edge of the town, and follow the crowd along a pavement so thick with snow it has been packed down by feet, rather than cleared aside by shovels. The central pedestrian walk is thronged with tourists, mostly Poles dressed in the latest ski fashions. Full of cafés, restaurants, high-street stores and helium balloons, Zakopane is like a crowded fairground, scented with cinnamon and mulled wine: a wonderland of temptations, but all a bit tacky. It's changed a lot since I was last here, in the summer of 1989.

We take shelter in a newly built wooden building with fresh log walls and suspiciously clean sheepskins on the pine chairs. But the coffee and scrambled eggs are good, and we look out onto a peculiar scene of men dressed in traditional costume beside horses and sleds, trying to drum up custom for a sleigh ride through the outskirts. The long, mournful faces of the horses are buried deep in oat sacks. The men are from another age, shorter than the tourists, with moustaches and leather waistcoats. Diesel-powered cars edge around them, looking for parking places. The horses stamp and snort with indignation, then return to their nose-bags.

The road to the mountains leads beside the Strążyska stream, past wooden homes built in the 'Zakopane style'. In front of us the peaks of Turnia Kiernia and Sarnia Skała are lost in cloud.

Staszic and his fellow enthusiasts 'discovered' the Tatras again and again in the nineteenth century. The locals were known as *Gorale* (highlanders), from the word *góra*, meaning mountain. Their loyalty was to their mountain villages, to their region of Podhale, with its jagged peaks and constantly changing weather. It was a borderland of mixed identities, a refuge for poachers and fugitives from the long arm of the law. As the nineteenth century waned, and the prominent visitors from the plains came thick and fast, the highlanders basked in their new-found glory. The Gorale seemed like a race apart to Adolf Hitler, too, a century later. When Nazi armies overran Poland in 1939, the Gorale were treated better than other Poles – in the peculiar Nazi view, they were somehow proto-Aryans.

Father Józef Stolarczyk arrived in 1848, aged thirty-two, and set about banishing superstition (as he saw it) and reminding his new-found parishioners that their good fortune to be born in the mountains meant they were closer to Heaven.

His job was not always easy. Patrice Dabrowski – whose *The Carpathians: Discovering the Highlands of Poland and Ukraine* is a rich resource for this chapter – cites the geologist Ludwik Zejszner: 'The passionate as well as gregarious highlanders seemed to accept premarital sex, love affairs, and children out of wedlock as nothing out of the ordinary.'

Stolarczyk had much to say about this in his Zakopane sermons for the next forty-five years.

Tytus Chałubiński, a doctor from Warsaw, which was then in the Russian-controlled third of Poland, visited Zakopane in the summer of

1873 to plant an iron cross on the Gubałówka hill overlooking the village. When the cholera epidemic struck, he stayed on for five weeks treating the sick.

Over the coming decades, Chałubiński spent several months here each year, and brought in patients from all over Poland. Zakopane's fame spread as a place not only of great natural beauty, but whose air alone could cure tuberculosis, one of the curses of the era. Three hundred paying guests visited the village in 1873. By 1883, there were 3,000, and the village was spreading up and down the valley.

In the same year that Chałubiński arrived, the Galician Tatra Society was founded to protect the Carpathian mountains, encourage tourism and protect rare animals like the chamois and the marmot.

Chałubiński launched what he called 'excursions without a programme' – trips lasting several days, with no planned itinerary or goal and often accompanied by musicians. Participants would carry their tents and musical instruments. The new hiking club built simple log cabins as accommodation and shelter. Highlanders in traditional costume led the hikes, mingling with middle-class doctors, lawyers, writers and artists from the plains. In a nation partitioned between foreign empires, Zakopane became a national symbol of unity, proof that Poland would one day be restored to the world map.

There's a strange parallel to the 'national excursions' organised by Ľudovít Štúr on the other side of the mountains. Štúr's excursions had fixed aims, whereas Chałubiński's were more hedonistic and closer to nature. And yet both were an expression of 'the nation' in a century when many intellectuals were wondering what a nation might be.

'During this trip in 1878,' wrote Patrice Dabrowski,

Jan Krzeptowski-Sabała (1809–94) played his squeaky and ancient fiddle for the group. The tales and music of this 'last old-fashioned highlander' seemed to transport all present back in time, longing for the primeval past that Sabala had brought to life.

We reach the car park at the beginning of the Strążyska valley, purchase entrance tickets and walk into the national park, still following the footpath beside the Strążyska stream. The water tumbles down the valley like children down the steps of a kindergarten, with a happy gurgling and chatter.

The morning mist has thinned to almost nothing, and suddenly the sun breaks through. There are quite a few walkers, but nothing like the crowds we left behind in the centre of town. Everyone seems bewitched by the sunlight. Overloaded spruce shed their loads, to the hilarity of those walking underneath, especially the children. The sunlight turns even the darkest conifers a brighter shade of green. The peaks of the mountains float in and out of sight between the clouds. Strangers greet each other like old friends. A grumpy Hungarian woman in uncomfortable high boots and a snow-white fur coat stands out. Her squeaky-voiced companion is trying to find the right angle to photograph her from. There's another signboard, warning of bears:

> The bear is a large predator which may be dangerous. The way you behave is crucial both for your safety and the future of the bear . . . Never approach a bear if you see one. Do not throw food – the bear might demand more. Do not panic – bears do not hunt for people. Just walk away calmly.

We reach the timber cabin that serves as a bar for thirsty hikers. Stupidly, I ask for a hot beer, rather than the hot wine I had set my heart on. I get a half pint of something half-fizzy of uncertain taste in a leaking paper cup.

From the cabin it's just a fifteen-minute walk to the Siklawica waterfall. At the foot of a slippery rock on the path, we stumble on the Hungarians again. She is complaining about the steep slope. He gives up on her and storms off ahead. The waterfall is a single thread of white water, like a white line on the dark-black granite, pouring down from the forest above. There's only a narrow, snowy platform to stand on, with the waterfall behind one's shoulder. Each couple or family cluster waits patiently in turn, though some bold photographers step out onto the rocks mid-stream for a different angle.

Up ahead, the highest Polish mountain in the Tatras, Rysy, with its three peaks, is out of reach. On the website of the Tatra Volunteer Mountain Rescue service there are fine views from a web camera, looking across the Morskie Oko – the 'eye of the sea' – lake towards the mountain. You can watch the sun come out, fleetingly, then the mist closes in, until visibility is almost zero.

Kazimierz Przerwa-Tetmajer was a Polish writer, born in the village of Ludźmierz in 1865, eight years before Chałubiński first came to Zakopane on holiday. He studied philosophy in Kraków and Heidelberg, and joined the Young Poland movement in the early 1900s. The movement, centred in Kraków, tried 'to revive the unfettered expression of feeling and imagination in Polish literature'. Tetmajer's best-known work was *Na skalnym Podhalu* (1910), published in English as *Tales of the Tatras*. The twelve stories have titles like 'Zwyrtala the Fiddler', 'Far-Off Marysia' and 'The Winter Maidens'. The famous fiddler Zwyrtala dies and goes up to Heaven, where he causes chaos by teaching the angelic choir wicked tunes on his violin. So he's sent back down the Milky Way to the Tatras, where he's still fiddling to this day.

In Stanisław Wyspiański's play *The Wedding* (1901), a poet marries a peasant girl. This was a common occurrence, as the Polish intelligentsia flocked to the Carpathians around Zakopane. A word was even coined for it – *chłopomaństwo*, which means something like 'peasant mania'.

In the play, after the wedding in Kraków, the guests move to the bride's house in the village for the party. Amidst much drinking, feasting and general hilarity, various guests from Polish history appear – the guilty consciences of the living, who are having fun while Poland, partitioned between the great powers, withers on the vine. Among them is Wernyhora, an eighteenth-century bard and prophet of Poland's re-awakening. He presents the poet-bridegroom with a magical golden horn, with which he should summon the people to unite against their oppressors. A farmhand is sent out to sound the horn at the four corners of historic Poland, but he loses it on the way. Meanwhile, the guests just keep dancing. The chance to save the country is squandered.

The play helped inspire Wyspiański's countryfolk. In December 1918, the pianist Ignacy Paderewski sparked an uprising with a patriotic speech in the city of Poznań. In June 1919, under the Treaty of Versailles, Poland was recreated, to include parts of West Prussia in the north and Silesia and the Silesian Beskid mountains in the southwest.

Tetmajer struggled with mental illness in his old age, and was sent to a mental hospital in Warsaw. The Nazi occupiers evicted the patients, and 'his starved, lifeless 75-year-old body was found among the ruins of Warsaw one witlessly cold afternoon' in January 1940.

In 1972, the Polish film director Andrzej Wajda made a film based on *The Wedding*. In the closing scene, the villagers stand or kneel stock still, as if bewitched, beneath pollarded willows in the mist. The magic horn has disappeared from the hay stook, to which the peasant lad had tied it. Only the string remains. The poet watches his white horse lie down in the mud.

'Jesus, the rooster has crowed!' cries the desperate, hungover farmhand. 'Hey brothers! On your horses! Take your weapons! The court at Wawel [castle] awaits you!' But the wedding guests remain bewitched. 'They hear nothing. Only the music. They all fell asleep.'

Then the people seem to wake, and their lances fall to the ground with a clatter. They start to dance, a slow, circular dance in the freezing mist. The chorus, repeats, over and over: 'Only the string is left.'

The town museum in Zakopane has just reopened. There are black-and-white photographs from the glory days of the 'excursions without purpose' – hikers idling beside their horses, gazing out over magnificent views of the mountains.

The contribution of the lowlanders to the highlanders' welfare was not just from tourism. Chałubiński provided clover seeds, green manure to help the highlanders improve their soil. He set up a bank to offer loans ahead of the harvest; and woodworking and lacemaking schools to provide skills for local people.

Five narrow valleys to the east of the Strążyska stream, a road leads south to Kuźnice. The discovery of iron in the seventeenth century offered a very different fate for the mountains. For a hundred years, until partition in 1795, pig iron was produced in the Hamerski foundry. The limitless supply of beech trees provided charcoal to fuel the furnaces. At their peak, the mines provided work for around 120 men, including miners from Germany and Hungary. By the mid-nineteenth century, the mine was no longer profitable and production ceased. With Zakopane's new-found fame as a visitor destination, tourism offered alternative, healthier employment.

CHAPTER 7

EVERYWHERE A FOREIGNER,
NOWHERE A STRANGER

In the west, the armies were too big for the land; in the east, the land
was too big for the armies.

Winston Churchill, *The Unknown War:
The Eastern Front, 1914–1917*

Poland, thought Lucius. *Galicia.* Somewhere, in the woods, they
must have crossed the border. On a table, inexplicably, was a beautiful ceramic music box, which played an unfamiliar tune. The bed
had been lanced open, and emptied of its straw.

Daniel Mason, *The Winter Soldier*

If you study a relief map of the western Carpathians long enough,
faces start to appear. One is of a Neanderthal man, with the High
Tatras for his flared nose, the Low Tatras as the upper lip, the lakes
either side of Nowy Targ in Poland as his eyes, and the gentle curves of
the Beskid mountains framing his face. The rivers of Poland are his long
hair, flow down his neck and away, all the way to the Baltic Sea.

The Vistula is the queen of these rivers, more than 1,000 km long,
rising on the slopes of Barania Góra in Silesia, close to the triple
border of Poland, the Czech Republic and Slovakia. Kraków is the first

major city on her path, but she also claims Warsaw, Płock, Toruń and Bydgoszcz on her way, before disgorging into the sea near Gdańsk.

The San is the Vistula's younger sister, rising in the eastern Beskids right on the border with Ukraine, and meandering northwest to reach the Vistula at Sandomierz. The Ukrainian national anthem, adopted in 1991, celebrates Ukrainian territory 'from the San to the Don' rivers:

> Brethren, stand together in a bloody fight, from the Sian [San] to the Don,
> We will not allow others to rule our native land.
> The Black Sea will smile and grandfather Dnipro [Dnieper] will rejoice,
> For in our own Ukraine good fortune shall flourish once again.

'Some memory of a liturgy, of an all-night vigil, lies submerged in this anthem. It seems as if the wind is blowing through this simple chant, as if the branches of a tree are singing,' wrote the modern Ukrainian composer Valentin Silvestrov. Silvestrov participated in the Euromaidan protests against Russian influence in 2013, where the anthem was constantly sung.

The Polish national anthem also draws on its own river, but in a slightly different context:

> We'll cross the Vistula, we'll cross the Warta,
> We shall be Polish.
> Bonaparte has shown us the way
> In which we shall prevail.

Napoleon means different things to different peoples in eastern Europe. In Poland, he's a rather positive figure. He established the Duchy of Warsaw in 1807 – the first attempt to recreate a Polish state after the disastrous partition of 1795 (though it was partitioned again between

Austria and Prussia in 1815). In Slovakia, he is disliked for his failed siege of Pressburg (Bratislava) in 1809 and the scorched-earth policy which followed – the destruction of the castles of the Little Carpathians.

On our first morning in Leśne Berdo, a guesthouse near the village of Przysłup, we climb through the wooded hills towards the ridge that forms the border. The snow has been falling since we crossed into Poland from Slovakia, and now lies half a metre deep, covering the railway track we walk beside as far as Przysłup station. It's a forest railway, not in everyday use, and the points stick out of the deep snow like tiny human figures. The sun is shining, a perfect day for the mountains, the temperature hovering around zero, just cold enough to keep the snow intact.

From Przysłup we follow a track up into the forest towards Jasło. Fir trees cast their shadows like giants in the winter sunlight. The snow puts polka dots on the silvery trunks of the beeches. It takes us three hours to Jasło, instead of two, slowed by the snow and the sheer beauty. The sun is still shining, but the wind is blowing white clouds out of Slovakia, while the clouds over Ukraine are darker. Geese fly in formation high above us. On the open mountainside, the wind makes delicate curves in the snow, like the marks on a sandy beach left by the sea.

The first peak we reach is Okrąglik, at 1,101 metres. The cube-shaped stone marking the border with Slovakia is painted white, with a coat of red round the top. Light seems to burrow out of the snow into the sky. We rarely meet another person, though there are footprints and animal tracks in the snow, mostly of deer and dogs.

Along a wooded ridge we walk, with Slovakia to the right of us, Poland to the left and Ukraine straight ahead. Our original plan was to cut down to Wetlina, then hitchhike back along the road to Przysłup. But we set out later than we had planned, and the winter's day is too short; so we come down to the valley at Smerek, and feast on trout and dumplings and excellent Polish beer.

The next day is much warmer, and the melting snow swells the San river at Lutowiska, only 2–3km from the Ukrainian border. The hills have no nationality, softly coated with beech trees, and everywhere the roar of the rivers. The wooden Greek Catholic church in Lutowiska was burnt down and the population expelled to the Soviet Union after the Second World War:

> The Second World War in Lutowiska effectively lasts 12 years: the local population is annihilated. The Jews are shot, Poles and Ruthenians deported. Lutowiska becomes Szewczenko. After the devastation and theft, the church has a new role – as a stable.

The Greek Catholic church of St Michael the Archangel in Smolnik survived, the oldest and most beautiful church in the Bieszczady region, built in 1791 of seasoned timber.

In 1951, this corner of Poland was swapped by the Soviet Union for a similar-sized plot, rich in coal seams. There was hardly anyone left alive. The landscape still seems strangely empty, with villages few and far between. This is a world deeply scarred by war and deportations.

In the First World War, this part of Galicia was a constant battleground between the armies of the Austro-Hungarian empire, to which Galicia then belonged, and Russia. Developments in Galicia had a profound impact on the outcome of the war. In August 1914, the chief of staff of the German army, Helmuth von Moltke, withdrew five divisions from his western front, advancing across Belgium, and sent them to Galicia. In the early stages of the war, this made tactical sense. The Germans destroyed two Russian armies, overcoming the overwhelming numerical superiority of the Russians with better strategy, weapons and supply chains. Their offensive on the Eastern Front in the summer of 1915 failed to deliver a decisive victory, however. In 1916, the Brusilov offensive was the last real effort of the war by Russian forces. Commanded by General Aleksei Brusilov, the Russians achieved initial success, then foundered on the combined defences of Germany and Austria-Hungary. By tying down German troops who could otherwise have fought the Allies at Verdun, events in Galicia played a crucial role in the eventual

Allied victory. Some 750,000 German soldiers, a million from the Austro-Hungarian armies and 2 million Russians died in the eastern Carpathians, and the plains stretching down from them to north, south and east.

According to military historian Timothy C. Dowling:

> Brusilov's armies regained all of the territory lost in 1915 and advanced once again to the Carpathian mountains, where they threatened Hungary. Only rapid action by Germany's military leaders held the front together and prevented the collapse of the Habsburg Empire in 1916.

Behind the big strategies and the numbers lay the vast misery of soldiers and civilians.

'From Szvidnik to the Dukla pass the road curved up rapidly,' wrote the American correspondent Alden Brooks in the *New York Times* in 1915.

> The forests were torn and burnt with shot and shell. Here were more scenes of violent struggle, trees fallen across the road, and now roughly shoved aside, rudimentary trenches, abandoned artillery caissons, two or three cannon with broken wheels, and soon, bloated, mangled horses, legs in the air, and finally the dead. They lay there to right and left, Russians for the most part, some dead in a last agony, their fingers clutching the air; others in the gutter where they had crawled to die, blackened face buried on an arm . . .

The American novelist Daniel Mason drums up no less bleak a scene in *The Winter Soldier*, but on the other side of the lines, among the Austro-Hungarian armies. Much of the action takes place in a makeshift field hospital in a ruined church in the (invented) hamlet of Lemnowicze. Margarete, a nun-turned-nurse, and Lucius, a trainee doctor from Vienna, extract bullets and shell fragments, perform amputations, and tend the sick and dying with insufficient medicines, equipment or food. The army, meanwhile, just wants to get the men back into battle:

And then there were others, men who could have fought again but now refused. Their war was over, they told him with finality. They had once been patriots, but all reasons for their patriotism had long been lost.

'Why should I shed blood for Austria? The Czech and Polish and Romanian and Hungarian and Ruthenian soldiers asked him. When Austria sends us into battle in front of her own? With shoes made of cardboard! And two men for every gun!'

'They will hang you for desertion,' Lucius told them.

'Ha! Let them come.'

In a bar on our last evening in Przysłup, I order a bottle of 'Śnieg na Beniowej' – 'Snow on Beniowa' – because of the beauty of the label. 'This was one of our first brews,' reads the entry on the brewery web page.

Originally planned as a Christmas beer, it quickly gained an empire of ardent fans and is one of the most frequently brewed nowadays ... Thanks to a special composition of different varieties of hops, it has a beautiful vanilla-orange aroma with a delicate note of ripe peaches, kiwi fruits and summer flowers. The taste is citrus fruits, cinnamon and nutmeg. The whole is crowned by a short, savoury bitterness.

In a place scarred forever by the bitterness of loss, the sweet bitterness of good beer is welcome.

Snow on Beniowa, just like rain in Cisna is strongly associated with the region where we live, work, make our beers and pay taxes. Beniowa is an extremely charming, displaced village on the border with Ukraine, on the southernmost tip of Poland. Famous for its beautiful, old linden tree growing in the middle of an old village. Richard Kaja, a famous poster artist, graphic designer and traveler, masterfully depicted it on the label . . .

Ursa Major is a small craft brewery in the village of Uherce Mineralne. The pub also serves beer from Ukraine – Lvivske – with the year 1715 on the label, which I also taste. But the beer from the small Polish brewery is undoubtedly better.

The brewery's first Facebook entry for 2024 applauds the decision of the new Polish government, elected in October 2023, to end the cutting of trees in the national parks and do more to protect the Carpathians – the opposite of what the new Slovak government is doing.

'It's time to move chainsaws out of parts of Polish forests,' said Environment Minister Paulina Hennig-Kloska. 'We have decided to issue the first decision to limit and suspend logging of the most valuable forest areas in Poland.'

The new government coalition also pledged to expand Poland's national parks. The outgoing government did not have a good environmental record. 'Foreign organisations have been demanding influence over Polish forests at the European Commission – and the [Court of Justice of the European Union] has just granted them this right,' Tweeted former Prime Minister Beata Szydło. 'It is worth emphasising – this is not about "defending nature". This is about giving foreign organisations the right to block Polish decisions.'

It is a strange, but familiar argument, often repeated by nationalist governments across eastern Europe. They're *our* trees, and we'll cut down as many as we like. The same wind blows through Brussels that ruffles the leaves in the high Carpathians.

⊚ ⊚ ⊚

The Black Dunajec rises on the northern slopes of Wołowiec mountain, then flows eastwards, collecting all the streams along the Polish–Slovak border. At Červený Kláštor (Red Monastery), by now just the plain Dunajec, it forges through a spectacular gorge, much loved by rafters and cyclists, to reach the Vistula.

The red monastery is a rather gloomy place, enlivened by an excellent legend. The monks had a reputation for the healing power of their herbs, under the guidance of a monk called Cyprian, who studied alchemy, cosmology and botany, and who wrote his *Herbarium* between 1756 and 1775. The original copy is in the Slovak National Museum in Bratislava. According to legend, Cyprian was also determined to fly. He constructed a flying machine, and launched himself from the 'Three Crowns Hill'. At first he flew like a bird, and soared so high that an angel saw his reflection on the Morskie Oko – the 'eye of the sea' mountain lake in the Polish Tatras. She struck him with lightning and he crashed to the ground. A rock, in the shape of a monk, marks the place.

◎ ◎ ◎

Like the Dunajec, the Poprad rises south of the High Tatras, then flows round them to the Vistula. On the shore of the Poprad, in Strážky, there is an exhibition of the work of the artist László Mednyánszky (1852–1919) in the manor house where he lived. Mednyánszky came from a noble Hungarian family, with Slovak, Polish and French ancestry. His most striking landscapes are of the Tatra mountains and the Great Plain of Hungary near the River Tisza. He worked as a war correspondent on the Austro-Hungarian front lines in Galicia, was injured, barely survived the war, and died in Vienna in 1919.

The castle is white as a wedding cake, with square towers on each corner, topped by battlements. The walls are vanilla-white, thick enough to be defended, but their colour evokes peacetime and ice cream, rather than conflict. There's a sense of waiting – for the artist to return, or Russian tanks to arrive.

In 1968, the French director Alain Robbe-Grillet shot his film *The Man Who Lies* here. After the Second World War, a man with a split personality arrives in a small town where no one remembers him, claiming sometimes to be Jean, a hero of the resistance, sometimes Boris, a traitor. Three lonely women are drawn to his odd and contra-dictory stories. He sleeps with two, but Jean kills Boris, his alter ego, before he can sleep with the third.

Mednyánszky was a chess player, a game he could find wherever his travels took him. He got on well with peasants and aristocrats alike. The great love of his life, Bálint Kurdi, was a boatman from Vác on the Danube. Another close friend was Blažej Ladeczky, a young shepherd from Potvorice on the Váh river.

Mednyánszky's paintings from the Eastern Front are the bleakest. A figure in white – his arms folded behind his back, military style, his clothes almost the luminous white of the Strážky castle – inspects a huddle of hooded prisoners of war. Another, 'Before a Captured Russian Trench' (1915), shows two black crows perched on twisted fence-posts, and others flying above. The barbed wire, in grey oil paint, looks as soft as cobwebs.

Before leaving the castle, we walk down to the Poprad. The surface is so still, reflecting white and grey cumulus clouds on a dark-blue sky. It is hard to believe the river is even flowing. Oaks and willows lean out over the water, like Adonis, to admire their own reflections.

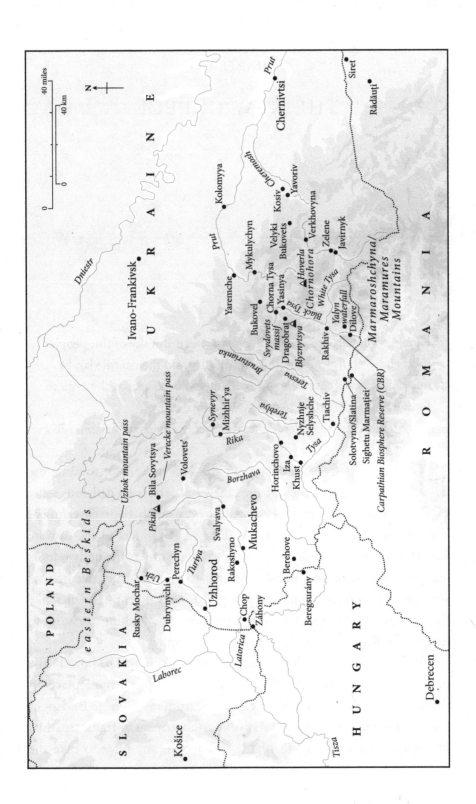

CHAPTER 8

THE NEW GREECE

Ukraine will become a new Greece: the beautiful sky of this people,
their cheerful disposition, their musicality and their fertile land . . .
one day they will awaken and its many wild peoples will merge to
form a cultured nation, just as Greece once did.

Johann Gottfried Herder, *Journal of My Travels in the Year 1769*

Heading east from Budapest on the M3 motorway towards the
border with Ukraine in June 2021, I experience a delicious sense
of illegality. Hungarians look west, however much their government
likes to trumpet 'an opening to the East'. The borders to the east are
harder to cross. The trucks belch more exhaust smoke, the body odours
are more powerful and the deodorants less subtle. But the light is
stronger, too, and the people less pampered. They don't put on airs, like
the citizens of European Union countries. I'm looking forward to
leaving the Roman, as well as the Austrian empire.

On the car-sharing app BlaBlaCar I find a ride to Uzhhorod. The
driver is a balding Italian called Alexander, who offers me a place on
condition I speak German. He picks me up from my home in Budapest.
He's just driven from Switzerland and has an apartment in Transcarpathia.
He's weary from driving all night, and accepts my offer to take the

wheel for a bit. The car feels unusually heavy and handles strangely on bends. It crosses my mind that he's smuggling something in the chassis.

Miraculously, there's no queue at the border into Ukraine, but a long snake of vehicles is coming the other way. This doesn't prevent the Ukrainian border police from giving us the once over. A scary blonde policewoman called Olena checks my passport and asks why I'm visiting her country. 'I'm a tourist,' I say, possibly for the first time in my life. I hate that word. She doesn't believe me anyway, sneers at my battered red British passport, the colour of a union my country no longer belongs to. She holds it disdainfully, like a blackcurrant stain in her palm.

'Proof of accommodation?' she barks.

In the 1980s, when I first visited the Soviet Union, you had to have such a document; but more than thirty years had passed since then. Hadn't she noticed? Fortunately, I have proof on my phone of two nights booked at the modest Hotel Atlant in Uzhhorod. A tall Ukrainian soldier pokes around in the boot of the car and tries to interrogate Alexander. 'What are these?' he demands, jabbing at two spare tyres in the boot, as if the wheel had not yet been invented. 'They're tyres,' replies Alexander, doe-faced.

If the soldier had a notebook, he would have written it down. Instead he mumbles the word into his walkie-talkie. Then he asks to see the registration number of the engine. This is more serious, but none of us can find it.

We poke around under the bonnet, among the hot, oily machinery that has just brought him halfway across Europe. Tiring of the game, the soldier loses interest and waves us through. Perhaps in Hell there's a special corner for border guards, where they can spend all eternity finding fault with one another's documents.

'This is all just *folklore*,' Alexander explains afterwards. The tall soldier and the blonde policewoman are just acting out the inspector and the interrogator. I wonder who they really are, and where the audience is seated. We're free to go.

For the last half hour along the dusty road to Uzhhorod, Alexander regales me, unbidden, with tales of his sexual exploits. Atlant Hotel? 'I fucked a lot of women there.'

We pass another, glass-fronted establishment. 'See that place? I fucked a lot of women there, too.' By this time, I'm relieved the journey will soon be over. 'You probably want to ask me, are there any places where I haven't fucked a lot of women?' he asks hopefully. Somehow I resist the temptation.

The Atlant is a three-storey, rather quaint establishment, with a friendly receptionist who tells me I'm the first foreign tourist this pandemic year. Then she asks about crossing conditions at the border. The border with Hungary sits on people's minds here. She's happy that I got in scot-free. A couple of years earlier, with her husband and small children, she crossed Hungary on their way to a seaside holiday in Albania. The Hungarians kept them waiting six hours in the sweltering heat. 'Imagine, with small children!' she says. More folklore.

I check into my clean, pleasant room, which looks out onto a cobbled street. There's a prominent neon sign above the entrance to a department store: KARPAT. The Cyrillic letters depict the mountains themselves – the vertical ascent of the K, then a difficult gully and hair-raising crossing in melting snow to the top of the A. The R offers many hand and foot holds, while the Cyrillic 'P' is like a table mountain.

I set off to explore Uzhhorod, the castle on the Uzh river, 115 metres above sea level. In the street near my hotel there are Gypsies, selling the first strawberries and cherries of summer, and bunches of purple-blue cornflowers. Beside them, an old man peddles an assortment of multi-coloured brushes. Long lines of linden trees offer shade. They were planted in 1928 to celebrate the tenth birthday of the Czechoslovak state: 'Linden avenues of freedom', they were called. The trees thrived within a root's reach of the river, and have provided shade and contemplation to the people of the city ever since. The avenue is 1km long.

Walking the streets, through columns of dust caught in the June sunlight, past old tin gates, with Soviet-era bric-a-brac for sale on the pavements, I am reminded how I fell in love with eastern Europe forty years earlier. Much has been lost in Hungary, beneath the new urban sprawl, the suburbanisation of everything. Not yet in Carpathian Ukraine.

I cross the footbridge over the Uzh. The river sits on its haunches, broad from the spring rains, polishing its stones. The Uzh is the first of the many parallel rivers I will cross on these western slopes, the fingers stretching down from the knuckles of the mountains. The Uzh, the Latorica, the Borzhava, the Rika, the Black and White Tisza, the Cheremosh and the Prut.

On the far side of the footbridge is a statue of the young Hungarian poet, Sándor Petőfi, hero of the 1848 Hungarian freedom fight. He is striding purposefully towards the footbridge and looks lonely, but determined. He penned six hundred poems in just eighteen months before and during the war of independence. At his feet there are wreaths in the Hungarian red, white and green. It's a good place for a statue: Joseph Stalin once frowned over the same square.

An armed guard patrols in front of the Hungarian consulate. The building was firebombed by Ukrainian nationalists in October 2018, angered by what they saw as the disloyalty of ethnic Hungarians to the Ukrainian national cause. The population of Uzhhorod was three-quarters Hungarian before the First World War, but only 7 per cent of the population call themselves Hungarian today.

'Who are you, actually?' I ask Pavlo Khudish, who has offered to show me round his city. Pavlo has read Paul Magocsi's *The People from Nowhere: An illustrated history of Carpatho-Rusyns* and even met Magocsi at a conference in the US. But he is not convinced by Magocsi's argument that the Ruthenes are a separate nationality: 'He adds Lemkos, Boykos and Hutsuls together and says they're Ruthenes. Yet they never thought of themselves as states or nations. This is a ball which gets kicked around. There's always someone who wants to heat up the issue.'

Pavlo's grandparents were Lemkos from Poland. Expelled to Ukraine after the Second World War, they made their first home in

Dnepropetrovsk (now known as Dnipro), then moved to Ternopil, where his mother was born. Then to Lviv, because they missed the hills. They ended up in a house just 20km from the Polish border. 'A lot of people identify with the Carpathian mountains, but don't like to be bracketed with a political or ethnic label.'

That evening we sit drinking beer under an ash tree, when a thin boy of about twelve years, with green eyes, appears at our table. Unlike the girls who toured the tables earlier, begging for small change, he's not from the Roma minority. 'I'm collecting money to make a film,' he announces, boldly. 'Would you like me to recite a poem for you?'

I ask what poems he knows. He offers a choice of six. Rocking back on his heels, he recites a short, romantic-sounding work of about four stanzas, his eyes fixed on the far bank of the river, as though on the audience in a theatre. It's by Lesya Ukrainka, the pen name of Larysa Petrivna Kosach-Kvitka, a popular Ukrainian poet and ethnographer of the early twentieth century.

'I would like to become a song,' Pavlo translates, '. . . to fly on the wings of the wind . . . to relish the waves of the sea, and the grasses on the steppe, and to be an echo in every heart.' The boy is paid generously for his rendition, and wanders off happily into the Uzhhorod night. Too late, I forget to ask what his film will be about.

Lesya Ukrainka was a poet, playwright and political activist. Her best-known work is a play called *The Forest Song* (1911). Mavka, a beautiful forest spirit, falls in love with a young man. The attraction is mutual, and she moves in with him, his ill-tempered mother and his uncle in a poor farmhouse on the edge of the woods.

Over time, the mother's hostility to the girl overcomes the love her son feels for her. Lonely and offended, Mavka rebukes the boy:

How pettishly you've broken off my rhyme!
Have you forgotten last summertime?
Of last year's summer I no memory keep
What was sung then died out in winter's sleep.

The heart-broken Mavka is driven out and replaced with a stout local peasant girl, whom the boy weds. But Mavka continues to haunt him. The uncle, who was more sympathetic to the forest spirits, dies and is buried under an oak. But the wicked mother has the tree cut down. The new wife, in turn, is driven out by the mother, and the boy, distraught, encounters the ghost of Mavka in the forest. He drowns himself in the marshes, and is reincarnated as a tree, reunited at last with Mavka, the two trees entwined.

Just behind the main square there is an orange-red synagogue, built in the Moorish style in 1904. The main entrance is shaped like a keyhole, like the Mesquita in Seville, a gateway to another world in Jewish and Islamic architecture. The synagogue is the Opera House now. All the Jews were taken from here to the death camps in 1942.

The castle, a chunky stone structure on a low hill just east of the city centre, is closed for the day, but I'm allowed into the inner courtyard to admire the linden and maple trees.

In a modern café and bakery near the city's second bus station, I meet Valentyn Voloshyn. He's running a workshop for artists, entitled 'Moving Borders'. One of the exhibits concerns a legendary tunnel which connected Transcarpathia to Czechoslovakia in communist times, through which cigarettes, drugs and weapons were smuggled. It began, he explains, in a wooden toilet in the basement of an old house near the border. There are also 3D mappers from Kyiv, who will project images onto the walls of the old town. The buildings, Valentyn explains, change the story each time.

◎ ◎ ◎

I leave Uzhhorod on a little green bus which bumps over the potholes, northwards, beside the Uzh. Ahead of us, the mountains shuffle forward, like bystanders at a demonstration.

The beech trees grow faster on the west-facing flanks of the Carpathians than anywhere else in the mountains. They love the westerly breeze, wafting up from the Great Hungarian Plain.

Beside me at the back of the bus, a young woman fixes her mascara and lipstick in a small mirror. She's very excited that I'm English – a real native speaker! She's an English teacher in a village further up the valley. Would I take part in their summer camp? The children have never met a real Englishman before! We exchange email addresses, then I'm off the crowded bus, bundling my rucksack apologetically over the passengers.

Alysa is waiting for me on the pavement, in smart walking boots and purple, pink and black hiking gear. I imagine Olha, the teacher in her pearls, gazing wistfully out of the window at my new companion, realising that I won't be there for her children.

An hour later, Alysa and I are climbing the footpath towards the mountain shelter at Javirnyk, an hour and a half's walk away. Below us in the valley sprawls the Gypsy village of Rusky Mochar – 'Ruthenian marsh'. 'I met a man there once who spent €18,000 on a horse,' says Alysa.

The steep path is flanked by beech trees and occasional spruce. Bicy, Alysa's energetic vizsla – a sort of Hungarian pointer – runs ahead of us. Alysa's husband Oleg went to Germany to work as a truck driver. Before that, he volunteered in the 'first' war with the Russians, in 2014.

Alysa and her friends are designing a 400km long-distance footpath through the Ukrainian Carpathians, from the border with Slovakia to the border with Romania. They have already walked the first section and plan to complete the rest of the walk by September. There are seventy or so enthusiastic hikers in all, including several children.

The Polish border is close by to the west, and beyond that the San river valley, Przysłup and Wetlina, where I walked in the deep snow the previous winter. After a steady climb through the woods, we reach a clearing. Ranged along the horizon to the north, I catch a glimpse of the main Carpathian ridge, above the Uzhok pass. 'The people here,' Alysa says, 'just want to make money from wood. They don't understand the soft approach to nature, the value of biodiversity. Punkt.' We're talking in German, and she uses the German word for full stop.

Near Pod Javirnyk, a carpet of pale purple flowers is spread beneath the beech trees. Five-leaved bittercress – *Cardamine pentaphyllos.*

The hut at Javirnyk is a substantial, three-storey house with a steep sloping roof, built on a foundation of stone and wood. Several young people are cleaning the dormitories upstairs for the next guests, cooking a hearty vegetable soup in the kitchen. Oleksandr Bursanov, from the Ukrainian Hikers Association, is in charge. His association maintains the huts, the footpaths and the signs on the trees. After a bowl of soup and a strong drink, we study the guest book, going back to the Second World War period, when Transcarpathia was briefly back in Hungarian hands. Many of the entries are in Hungarian.

Near the hut are several ancient stands of beech trees, the legacy of Alois Zlatník, a Czech botanist who worked in the eastern Carpathians in the 1930s. He marked out plots of virgin forest, and painstakingly noted the details of each tree in Czech and German. The maps are beautiful, like an architect's drawings. Zlatník was a keen ecologist, aware that the forests could be lost to intensive logging. According to a 2021 study, nineteen of the forty areas he mapped have been destroyed; eighteen are still more or less intact.

I sleep soundly that night in the converted barn in Dubrynychi. In the morning, Alysa picks me up and we drive down to Uzhhorod together, for a two-day hike along the Transcarpathian ridge – the next leg of their footpath. There's Oleksandr, who runs a bicycle shop in Uzhhorod; his wife Gabriela and their eldest daughter Solomiya; Serhii a mountain guide; Viktor Stenich, who runs a travel bureau; Gyuri – a Hungarian . . . Twenty people in all.

First stop in the van, a rather smart café in Svalyava. The owner will walk with us, and insists that we eat so many cakes, and down so many shots of various berry brandies before we set out, that I feel more like curling up under a tree than climbing a mountain.

By the time we start climbing from Roztoka, it's early afternoon. The path climbs steeply through a sea of wild garlic beneath a beechwood. Once we reach the ridge, we will be walking on the flat, or going downhill again. The rain sweeps in and I'm ill-prepared, my trousers drenched from ankle to knee, while the others have gaiters to protect them from the long grass, and big capes so the rain runs straight off them. My heavy wintry jacket is good against the cold, but an added burden in the sun.

Solomiya is only nine, but leads the way with Bicy, Alysa's dog, who covers at least three times as much ground as the humans in the party, sprinting forward and back.

Above the treeline, the open meadow is almost as thick with bilberry bushes as the woods were with wild garlic – but not yet ripe, alas. There's a fine view westwards, towards the Turiya valley, which flows into the Uzh near Perechyn, and the border with Slovakia. To our left, the pass at Uzhok is not far away, but we turn right, towards the Pikui peak, our main waymark today.

One moment the sky is clear, with views over the far side, eastwards; the next moment a thick mist rolls in waves, obscuring the path ahead, a narrow, rocky track. I stick with those with whom I have a language in common – Alysa in German, Gyuri and Gabriela in Hungarian, Taras, Viktor and Oleksandr in English.

Viktor used to work for the Swedish furniture giant IKEA, sourcing wood for their furniture. Then he was manager in a local charcoal works, 'supplying top-quality charcoal to the far ends of Europe'. Then he ran travel webinars, and finally set up a small travel agency of his own, encouraging Ukrainians to stretch their wings. The coronavirus pandemic made that rather challenging. 'Each of the companies I worked for was smaller than the last,' he smiles.

The beauty of the long-distance path is that it should be self-sustaining. Businesses along the road will advertise their food or accommodation. A booklet has been produced, which you can stamp on each completed segment.

The first day we walk 27km. On the summit at Pikui, an old metal cross stands in a block of concrete. Someone lets off a blue and yellow smoking firework – the colours of the Ukrainian flag. Meanwhile Alysa, who is responsible for song and dance in the group, leads a crazy jig, and we sing a ballad from the top of the Ukrainian pop charts. It's raining again, and we all look rather bedraggled in the group photograph.

Solomiya settles down beside me in the heather, over lunch, to show me her drawing book, full of bright-coloured birds, rabbits and squirrels, mums and dads and daughters. Then the sun comes out.

We finally reach our destination after nine in the evening – a small hamlet called Bila Sovytsya, 'the white owl'. These Ukrainians seem to mysteriously agree routes and resting points without any discussion. The village is quiet and we eat quickly and sleep deeply, lulled by a mountain stream. Our supper trout come from a pond just up the hill.

The next day we make an early start, walking among chocolate-brown Carpathian cows, the *bura karpatska*. Serhii says their front legs are slightly shorter than those at the back, to make it easier to go up the mountain. I'm not sure if he's joking.

Serhii has a penchant for striding ahead. He's a tough, muscular fellow, with great knowledge of the mountains – the kind of man you want beside you on a trip into the unknown. He also has a reputation, murmured among our company, that he is a *mol'far*, a storm wizard. When a hailstorm struck on some exposed crag on a previous hike, Serhii swept off his green hunter's hat and held it out. The hail stopped immediately. There are tiny irises, yellow, white and purple in the grass. And fresh purple cones, raspberry red, taking shape on the spruce.

There's a lot of magic left over in the Carpathians which doesn't fit into the sturdy towns of the plains, with their sleepy streets and ancient connections to the Hungarian plains beyond the Tisza. The mountains were always a place where people laboured – with sheep, or cows or wood – or else fled to for safety. The trees don't whisper here, they roar in the wind. If the wind is coming from the right direction, they say, the *trembita* – shepherd's trumpet – can be heard 10km away.

The footpaths we follow are well signposted, with red, blue or yellow waymarkers on trees or rocks, and distances measured in hours rather than kilometres. Alysa, Viktor, Serhii and their friends are just trying to connect them together.

We reach Verecke, at 841 metres – the mountain pass where, according to 'Anonymous', a twelfth-century author of dubious veracity, the Hungarian tribes crossed the Carpathians in the year 897. It seems as good a place as any. A Hungarian sculptor from Mukachevo, Péter Matl, designed a tall, stylised stone gateway – 'between East and West' – composed of three overlapping blocks on either side, with a stone triangle joining them at the top – a pagan symbol that is surprisingly pleasing to the eye. A sort of altar has been placed beneath it, draped with ribbons and wreaths by passing Hungarians, and a stone staircase leads down the hillside, to the west. 'What does this place mean to you?' I ask Gabi, one of the two Hungarians present. 'Not a lot,' she admits, ruefully.

In the thirteenth century, the Tatars rode through here to pillage the Hungarian kingdom. Just down the road, there's a sad graveyard and memorial to 600 Ukrainian Sich partisans allegedly executed by Hungarian and Polish troops in March 1939, though the details are still disputed. 'No order of execution by anyone affiliated with the Polish state authorities has been confirmed by any known historical source,' wrote Łukasz Jasina on the Polish Institute of International Affairs website.

Verecke is a symbol of the historical grievances between the different peoples of the Carpathians, rather than a gateway through which they all pass. Looking to the east, there are still traces of snow on the Chornohora – the Black mountains.

We walk slowly and wearily down to Volovets, where the vans are waiting. After two days' walking, I'm just getting the hang of a long-distance lope, and feel a little jealous of my companions, who will continue this journey soon, all the way to Romania.

◎ ◎ ◎

Public transport by small bus in Transcarpathia is efficient and extremely cheap for my Hungarian wallet. From Mukachevo, I take the small bus to Khust.

Oreste Del Sol sweeps me up at the bus station, and takes me straight to the Valley of the Daffodils, to the north of the town, on the edge of his village, Nyzhnje Selyshche. There are not many flowers left, he warns me, but we might be lucky. The wild daffodils bloom in May each year, and are one of the famous sights of the region – a great expanse of white flowers with yellow centres. They're a bit smaller than the daffodils of my youth, on the borderlands of England and Wales in Shropshire, but no less beautiful. We are rewarded with the last few flowers; but the purple irises, dotted through the meadow among them, are even more beautiful. We drive on up the hill, past Gypsy children offering bunches of the last wildflowers they have managed to find, and punnets of little wild strawberries that it must have taken them hours to collect. Their mouths are smudged red by all the strawberries they have eaten on the way.

Oreste came to Ukraine in 1992, aged twenty-four, when it was still in the Soviet Union. He belongs to Longo Maï, a cooperative of peaceful anarchists who began to buy land and establish communal ways of living in the aftermath of the student revolution in France in 1968. According to the Longo Maï website:

> In the big cities they could not realise their dream of a life lived in
> solidarity with one another, and they went in search of open spaces
> in the mountain ranges of Europe suffering from the rural exodus.

In the vernacular of the Provence region of southern France, 'Longo Maï' means 'Que ça dure longtemps!' – 'May this last a long time.' Fifty years seems pretty good going. There are communities in France, Switzerland, Germany, Austria, Ukraine and Costa Rica. They tend the land, ecologically, for the movement and their children. There are around 200 adults in the movement today. Oreste met his wife Jolana here.

The farm is on the edge of the village, beside a wood of handsome, century-old beech trees. Oreste is stern and humorous at the same time, and Jolana thin and friendly, rushing round the kitchen feeding dogs, cats and stray people, just in from dealing with the cows, pigs and goats. She's from the Hungarian minority in Ukraine. Oreste was born in Paris, in the Rue des Martyrs. They have a daughter aged twenty-six living in Basel, and a son of twenty-two in Paris. A large Bernese mountain dog, Stella, fusses around her four puppies.

For supper there are omelettes with goat's cheese, homemade bread, bulgur wheat and different salads from the garden, washed down with a good red wine from Berehove. Oreste is especially proud of their compost toilet. I sleep that night upstairs in their son's bedroom.

The next day we look around the cheese factory, where milk from cows from the neighbouring villages – 150 producers in all – is turned into three different types of cheese. Before he arrived here, most of the cheese available was from sheep – the soft white, feta-like cheese known as *bryndza* from Slovakia to Romania.

More than half the shepherds have disappeared from Transcarpathia in the past ten years, Oreste says. With his help, those left have set up an association of their own. Oreste has a strong desire for social justice, from his background in Longo Maï. 'To be a shepherd in Ukraine is actually illegal, or a-legal,' he explains. 'There are no relevant laws, no subsidies whatsoever, and once you have produced your cheese, you have to sell it illegally, in an undocumented way. No one even asks for veterinary documents, to find out the health of your sheep.'

Nevertheless, this is paradise, Oreste maintains, compared to the life of a hill farmer in the French Alps, with all their subsidies. They are driven to suicide by their dependence on the state, and by the bureau-cracy that entails. In France in 2017, a farmer committed suicide every two days.

Farmers in France feel hopeless, the organisation Paysans Solidaires told a reporter:

Because they cannot pay back bank loans or the vet's bill. Why is that happening? Put simply, the price they can sell their produce for is not enough to cover their expenses. And so they work harder, hoping to earn more money. But then they become exhausted. And the debts continue to pile up . . .

The answer everywhere, Oreste maintains, is not more subsidies, but more expensive food. People should get used to paying the real price:

> It's important for us to show that all this work by hand is still useful, so if tomorrow petrol becomes too expensive, we can return to this technique by hand, and food products will take on a new value . . . The problem today is we give more importance to the production of a smartphone than to agricultural produce. If farmers are properly paid, people can return to the land, and make the countryside and all its biodiversity come alive.

The cheeses are round and magnificent, stacked eight shelves high, each weighing a kilo. They are expertly handled by a lady in a turquoise apron and purple gloves. The cheesery is modelled on cheese production in a Swiss village. In the Soviet era, cheese factories processed 100,000 litres of milk a day. This one handles just 1,000 litres a day, from which 200 half-kilogram cheeses are made each day. It was built up over eight years, from 1994 to 2002. Very little is mechanised.

> At the beginning, they didn't believe that the milk brought in by 150 peasants could create a quality cheese. The EU has a tendency to industrialise things, and get rid of manual work. Here we only have manual treatment, which increases the quality.

The other factor which makes the cheese so good is the quality of the grass the cows graze, and the hay in winter, rich with wildflowers.

The cheese is sold in the farm shop for €12 a kilo, almost double the price in the supermarkets – true to Oreste's principle that food should be expensive. But they still manage to sell it all, because it is simply better than anyone else's.

Oreste also found investors to develop a new sewerage system for the village, and to improve the fresh water supply from a spring in the hills, 4km away. The sewage plant was needed to clean the acidic water coming from the cheese-making. 'It took so long because we stuck to our principles – no bribes for local officials to oil the wheels.'

Photographs of cheese-making in Nyzhnje Selyshche through the ages line the brightly lit, brick-walled corridors that lead from the cheesery to the restaurant. A dark-faced lad in a white coat leans on two enormous cheeses, cut to reveal their Emmental-like holes. He wears a tight, sheepskin hat.

Upstairs, I eat lunch in the restaurant with Oreste – pickles, marinated trout, black local mushrooms and a rocket salad, followed by a selection of cheeses of different ages, decorated with walnuts. Everything is beautifully served – like a work of art from the best French restaurant. There is even a little hill of pale orange trout caviar. We wash it down with wine and *uzvar*, a traditional drink made with dried fruits – apples, plums, pears and raisins, boiled briefly in water, left to stand and then drunk cold.

We drive together to Svydovets, a massif in the east of Transcarpathia. Oreste is leading a campaign to save the unspoilt mountain from developers, who want to turn it into one of Europe's biggest ski resorts. 'Svydovets is unique. There are very few massifs like it left in Europe, which are almost untouched by man. Some parts are already protected. But that's not enough.'

More than 50 per cent of the land in Transcarpathia is covered by forests. Ski slopes, and the infrastructure of roads and buildings they bring with them, are a big problem in the Alps. He's fighting this project both to save the massif and for a bigger, environmental goal:

We understand now that the climate is changing. We're waiting for the collapse of the world as we know it, when there won't be enough petrol. We have to change something in our way of life, for the next generations. We won't find enough energy, and human beings will despair. We have to adapt ourselves to nature, and not adapt nature to us.

On the outskirts of Dilove there's a blue-and-white monument at the roadside, dating from 1887: another claimant to the geographical centre of Europe.

We drive through Rakhiv, then branch north towards Yasinya. The road which was following the Tysa for so long, along the border with Romania, now follows the Black Tysa. I feel far from the Europe I know, hidden deep in the folds of the mountains. The Black Tysa is shallow and wide, hurrying over rocks. An ugly concrete embankment has been built in many places, against flooding, and to save the road from collapsing into the river.

In Yasinya, we transfer to what the Russians call a *bukhanka* – a 'loaf' – a chunky Soviet all-purpose vehicle, the UAZ-452. As we bounce the 12km up the mountain road to Dragobrat, it appears to have no suspension. In the spacious interior, we hang onto the roof rails for dear life.

Dragobrat is a small ski resort on the edge of Svydovets with about fifty hotels, which Oreste accepts are a necessary evil. This would be just the little toe of the giant complex which the Ukrainian–Israeli oligarch Ihor Kolomoisky and his business partner Gennady Bogolyubov are planning here. There will be 450 hotels, 230km of ski slopes, 120 restaurants and 23 ski lifts, making up one of the biggest ski resorts in Europe. The men already own another ski resort at Bukovel, 30km away. The Svydovets massif stretches for 25km, with adjacent spurs and valleys, towards Bukovel; so the plan is to link Dragobrat to Bukovel over a pristine mountain range.

The state forestry committee and the regional governor have given the project their blessing. Construction of a road through the forest has begun, from the village of Lopukhovo. An alliance of environmental groups is fighting them in the courts. They argue that pristine wilderness – including primeval forests – will be laid waste and the basin of the Black Tysa river destroyed. Also, the relatively low elevation (no mountains higher than 1,800 metres) at a time of climate change means there will be less snow, year by year. Even the reservoirs, which the project envisages will be built in high mountain areas in order to make artificial snow, will harm fragile local species of flowers and lichens. The final decision rests with the Ukrainian Supreme Court. The same Ukrainian oligarchs stand behind the existing Austrian ski resort at Semmering.

My hotel window looks east, with a fine view of the Chornohora range, topped by Ukraine's highest mountain, Hoverla, at 2,061 metres.

> Beneath the storm-snowed Hoverla hills
> hollers and hums a green mountain stream
> at midnight the fragile bridge trembles
> trampled by the feet of the naked maidens
> the broad love stream, roaring, racing
> lost in the stream's loud lamentation.

In the Hungarian original by Endre Ady, *Hóvár* means 'snow castle'; but lingering behind it, unspoken, is the word *vihar*, meaning 'storm'. The poet uses the image for his triple addiction to wine, women and words. He died of syphilis in Budapest in 1919.

Oreste and I spend the next day exploring the massif, partly with the help of our 'loaf'. We are high above the treeline. Snow drifts litter the north-facing slopes like lingerie. Where the snow has melted, purple crocuses are brilliant against the tattered grass and the grey-coated snow. There are also bright yellow mountain cowslips, known as bear's ears, *Primula auricula*, part of the primrose family.

The peaks stretch to the horizon. We clamber through dwarf juniper bushes to the Velykyi Kotel peak at 1,770 metres, where a rickety sign points to the Stig peak (1,704 metres) and Blyznytsya (1,881 metres). Far below us to the east, in the thick wooded valley, near the source of the Black Tysa, we see bald patches of clearcut forest. To the west, thick forests roll down the mountain to the Teresva and Brusturianka rivers, which curl round and through the mountains towards the Tysa.

The investors' agents have offered unlimited prosperity to the villagers, if the ski resort is built. 'Am I poor or rich?' Oreste asks, rhetorically.

If I live in ways that do not destroy nature, I'm a rich man. And if I destroy something, and maybe I have a Mercedes or a big house and I don't need to work any more, then I am poor, with a poor mind. It's hard to explain this because people want to have a better life. Life was always hard, here in the mountains. But that is because there was always some kind of imperialism – Austrian imperialism, Russian imperialism, then Soviet imperialism. We need more democracy here, then people will find out how to live. When governments tell me how I should live, I'm afraid that is not my way. It's just a mirage. A lot of people do not understand this, but we have to explain it. It's a matter of culture, and education, and the media. Maybe because I was born in the centre of Paris, with its bad air, its pollution, I'm looking for another life. People who were born here find life in a village difficult and closed. They work abroad, and they want to live like people in Spain or Germany. But I believe in the intelligence of people. In a few generations they will change their minds. The people who want to build here do a lot of things which are illegal. As they are very rich, they don't even know that the law is higher than their own power. The oligarchs have their own media, they have money, their loyal groups of parliamentary deputies, and they think they can do anything. Ranged against them, we have international public opinion. We cannot stop every project, but we will stop this one.

In May 2003, Ukraine and six other Carpathian countries signed the Carpathian Convention in Kyiv:

> By 2050, the Carpathians [will become] a thriving and sustainable region where people live in harmony with nature. The biodiversity and natural beauty of the Carpathians are conserved, restored, and wisely used, providing a healthy environment and essential ecosystems services for all people of the region and beyond.

The question is what happens until then.

Beside a dark-grey mountain lake fringed with skirts of snow, one of the 'eyes of the sea', some German tourists in a Range Rover have got stuck in the mud, after trying to drive through the lake. We watch them for a while, through a sudden squall of driving rain and hail which buffets the 'loaf'. We brew fresh coffee on a gas stove in the spacious back of the van. The Soviet technology suddenly seems far ahead of the sophisticated West.

Eventually, they get free and drive away over the mountain, while we bounce back down the mountain track to Yasinya. The mountains are rich in water and forests, two of the greatest sources of wealth in the twenty-first century. But the people here believe they are poor. If the oligarchs have their way, Oreste says, the people will remain poor. What is needed is sustainable, low-impact tourism, to develop the region in a way that taps the natural wealth of the mountains, without destroying it.

CHAPTER 9

THE BLACK TYSA

A yew root is everlasting. Leave a pasture untouched for a hundred years, with no one mowing it, and a yew forest will spring up again from under the ground.

Stanisław Vincenz, *On the High Uplands*

Yasinya is a small, pretty town of 9,000 inhabitants, nestled into the Black Tysa valley at the foot of Svydovets. From January to June 1919, it was the capital of the independent Hutsul republic; then Romanian troops marched in to end the country's brief independence. The end of the First World War was a time of new borders and new countries, and it was hard to tell which might survive the collapse of Tsarist Russia and the Austro-Hungarian empire. Yasinya became part of Czechoslovakia. For three days in March 1939, it tried again, as part of the independent state of Carpatho-Ukraine. This time, Hungarian troops ended the experiment, and ran Transcarpathia until the end of the Second World War, when it was incorporated into the Soviet Union.

On a sunny June morning, there's a blue-and-purple poster of Oleksandr Shevchenko, parliamentary deputy and ally of the oligarchs, on the facade of a hotel in the town centre. There are still traditional timber houses in the town, including the former Budapest Hotel – a fine, triangular structure with a long, sloping roof. Built in the early 1940s, it's now called the Edelweiss. Under some pine trees is the war memorial: a lone soldier in forest green, clutching his helmet in his left hand, his head bowed in thought. At his feet, on twelve crumbling, almost illegible plaques, are the names of the dead of two world wars. Along the ridge above the town stretched the defensive Hungarian Árpád Line. Outside the secondary school is a statue of Oleksa Borkaniuk, a communist partisan from Yasinya, who was parachuted back into Transcarpathia with a small group of Hungarian and Czech partisans in September 1941 to fight the Hungarian army. His main task was to sabotage the railways. Borkaniuk was captured in February 1942 and executed in Budapest.

I stay in Yasinya at a small guesthouse run by Svitlana, a friend of Oreste's. She walks with me along the shore of the Black Tysa towards Béla Franz's waste-recycling place. The threat of future destruction from the Svydovets project is one thing. Quite another is the mess the river is already in, due to human carelessness. 'The people just throw their rubbish in the river, expecting the waters to carry it away.'

On top of all the individual items – the PET bottles and plastic bags on the beach, caught up in the roots of trees; the broken beer bottles among the driftwood – the shocking fact is that people bring whole sacks of rubbish down to the shore. Another problem is the competition between many small companies to collect money from households to take away the waste. There is very little recycling. For the higher river valleys, the companies do not have appropriate vehicles, so there is no one to collect the waste. The river is brown from all the recent rains, and would be beautiful – a foaming mountain torrent, rushing through the centre of Yasinya – were it not for all the junk it carries.

Béla's waste-collection point is on the shore. He's a busy man, lugging sacks, speaking on the phone, meeting townsfolk with small bags of waste, or tradesmen coming to take the sorted waste away. An old bloke in a red jacket arrives, pushing a blue bike, onto which he has somehow stacked four white sacks of beer bottles – the result, he says, of six months' drinking with his friends.

Béla helps him put 120 bottles into yellow crates, then hands over a small wad of scruffy hryvnia notes – sixty in all – enough to buy four half-kilo loaves of bread. I help Béla stack the crates in his truck. Twice a week he does a run to the nearest city, Ivano-Frankivsk, where the bottles are processed. Three women sort through white and green plastic waste and cardboard. He has a pressing machine for the tins and a baling machine for the cardboard. He's indefatigable, running from room to room in the heat of summer. Svitlana and her friends volunteer here out of idealism, to help clear the valley of all the waste.

She shows me his 'museum', full of strange objects he has found, glass bottles in special shapes. He has a soft spot for Jägermeister bottles, she says.

Each summer, Attila Molnár and his team in Hungary organise a 'PET cup', sailing rafts made entirely of plastic bottles down the lower reaches of the Tisza river, to publicise the waste problem. The Hungarians are no angels, Attila says, but the worst culprits are upriver – the choking flood of bottles floating downstream on the Tisza from Ukraine and on the Someş from Romania.

When he finishes work, we drive together to Bukovel, to see the ski resort owned by Oleksandr Shevchenko. There are roadside advertisements for swanky hotels, a Dino park and erotic massage. Polyanytsya was a quiet village, turned into a mini-Las Vegas by the ski resort, which opened in 2002. Security guards loiter menacingly at the entrance to car parks and hotels. There are seventeen ski lifts and 60km of slopes. In the winter of 2010–11, there were 1.2 million visitors. In a country where there are very few opportunities to ski, Bukovel almost cornered the market.

We stop in a car park beside a storage lake, on the far side of Bukovel, watched by suspicious security personnel. The lake is to produce artificial snow. 'I remember when there were just a dozen or so houses in this village,' says Béla. 'Now the only locals employed here are guards or cleaners. They bring the waiters from Kyiv, and locals can't even gather mushrooms in the woods without permission from Shevchenko.' Leaving town, we pass the refurbished secondary school – one thing, at least, which the resort has contributed.

On the way back to Yasinya, many houses advertise the famous Ukrainian hot tub, the *chan*. These are made of cast iron, heated from beneath by open fires, with large stones on the bottom, scented with herbs – rather like being cooked up in a giant's cauldron. There is usually a second, wooden tub of freezing water into which you can leap when your body is hot enough from the *chan*.

The next day, with Svitlana and Béla, I set out to the source of the Black Tysa. 'If this project does not happen, nothing will get done here: no new roads, no new schools, no jobs, nothing,' said the mayor of Chorna Tysa, Ivan Pavlyuchuk, in a 2019 interview with Andrew Higgins of the *New York Times*.

Beyond his village, walking through the forest towards the source of the river, under the lee of the Svydovets massif, we come across large areas of clearcut forest. The trees were cut some time before, the stumps have been abandoned, and there is little sign of any regeneration.

'The forest belongs to the state forestry company, but most of the felling is done by private entrepreneurs,' explains Béla. 'Both blame the other for not tidying up after them, and not planting new trees.' Clearcutting is the cheapest and quickest way to fell a forest. The alternative would be selective cutting, which would leave trees of different ages to grow. A lot of wood is cut illegally. 'The deal is this: you cut the wood where you like, then split the proceeds fifty-fifty with the person from the state authority who is supposed to be protecting that segment of forest.' Those who cut the trees earn ten times as much as local people – 'and change their cars every year, unlike the poor locals, who can hardly afford a bicycle'.

In areas of forest that are still intact, the forest floor is rich with wild-flowers. Deep-blue snow gentian, *Gentiana nivalis*, different species of purple iris, crocus and golden archangel, *Lamium galeobdolon*. Through the spruce trees, we catch glimpses of snow patches on the flanks of Svydovets, like straggling teenagers on a school outing.

Closer to the source, the forest seems better protected, the soft sunlight falling through tall, undisturbed trees – pine and spruce, 30–40 metres high. The source of the river itself is a stone wall, out of which a steel pipe pours the precious water into a round receptacle, from which it flows on to become a stream. The area all around is paved. In a keyhole arch around the source, the blue-and-yellow Ukrainian colours have been painted on the stone. The ambassadors from the countries on the shores of the Tisza and the Danube planted a yew tree to celebrate the cooperation between their countries. The word for 'yew' is *tys* in Ukrainian and *tisza* in Hungarian: from these words the river gets its name(s). Here, at least, we can drink the sweet, unpolluted Tisza waters, at 1,680 metres above sea level.

Perched on a rock, Béla tells me his story. His great-grandfather was brought here as a forester from Hungary in 1918. His grandmother was born in 1924. He has worked in Hungary and the Czech Republic, but was always drawn home by this landscape. One day he was swimming with his son in the Tisza, when a plastic bag in the water caught round his body. 'That was it,' he says. He went to Hungary, learnt the trade, then came back to start his own waste business. Everything is sorted by hand, but in future he wants an automatic waste-sorting plant.

The town council has already offered a piece of land. Next he's looking for investors to provide the machinery, including small trucks able to climb these steep hills. He would simply be the manager. He just needs half a million euros.

Another part of the solution would be to reorganise and develop the final destination of the waste. Above all, the mentality of the people needs to change. 'Throwing organic waste into the rubbish bin is a crime,' he says, bluntly. 'Everyone here has a small garden or plot of land, where they could compost it.'

Private rubbish companies come from Ivano-Frankivsk, and people pay them by the cubic metre to take their waste away – 'they can't even be bothered to squash the bottles, and as they're paying by volume, they're paying for air!' he fumes. Only about 25 per cent of waste cannot be recycled, and that should be buried or burnt.

On the way back, we stop at a forester's cabin to make an outdoor fire, starting with a pile of fresh wood chips on a cardboard sheet, because the fireplace is so damp. We feast on small pies stuffed with sheep's cheese, made by Béla's wife and his daughter Emma, washed down with *uzvar* made from blueberries. Driving back to Yasinya, the mud caused by the rain has cleared, and for the first time I see the clear waters of the Black Tysa, crystalline and slate grey.

◎ ◎ ◎

The headquarters of the Carpathian Biosphere Reserve (CBR) is in Rakhiv. 'This is nature undisturbed by humans,' boasts its website, 'the world's largest contiguous primeval beech forest area . . . the place where 500-year-old and 45-metre-high beeches grow.'

Vasyl Pokynchereda and Iryna Yonash from the CBR take me for a long hike, up into the woods above Rakhiv. Bats are Vasyl's speciality, in the caves hidden deep beneath these mountains; but he also knows a lot about trees.

On the CBR map, the ancient beech forests, recognised as a world heritage site by UNESCO, are marked in brown; the massifs in olive green, including all of Svydovets; while by far the largest areas are mustard yellow. These are the 'transition areas of the reserve'. Like 'buffer zone', 'transition areas' leaves a little too much to the imagination of those who want to cut the trees.

Vasyl remembers as a child watching rafts of logs, steered by foresters, coming down the Cheremosh, one of the wildest and most romantic of the Transcarpathian rivers, draining the slopes of the eastern Carpathians. 'When the first raft got stuck on a bend, the men on it had to scramble for safety, before the next raft slammed into it,' Vasyl remembers. 'They fled like mice!' When a raft of timber broke up, it could take several men a whole week to tie it back together again for the journey downstream.

Beside the path we come to an enormous silver fir, *Abies alba*, straight as an arrow aimed at the sky, perhaps 300 years old. 'One peculiarity of the silver fir, even after they die from natural causes, is that they can continue to stand for up to 100 years!'

The beech rots quickly and normally collapses within a decade. And both, as they rot, provide a perfect breeding ground for all manner of beetles, insects, fungi and lichen, which form the rich diversity of the forest.

'Do the forest managers in Ukraine understand the value of dead wood?' I ask, fresh from my conversations in Slovakia.

'It's a beautiful question!' Vasyl replies. 'The old Soviet school of thought is still dominant here. According to that, the less dead wood in the forest the better.'

Things are changing, however. Young foresters learn new approaches to forestry in western Europe, where there are no virgin forests left. But, he laments, it will take decades before appreciation for dead wood is really established in Ukrainian forestry theory and practice.

Another big problem is corruption. One special form this takes in Transcarpathia is for the sanitary inspectors of the forests to declare healthy trees sick, so that they can be felled in so-called 'sanitation logging'. Some inspectors get rich, and the forests are decimated.

We climb up through a cathedral of trees, all the way to the magnificent Yalyn waterfall. There's no one else there, no sign of human disturbance in the forest; just the roar of the water and the fine spray constantly soaking the trees.

◎ ◎ ◎

We meet Vasyl Andraschuk in the restaurant of the Hotel Europa in Rakhiv. 'I am a Ruthenian, but also a Hutsul. Being Hutsul is a big part of my identity. I don't feel Ukrainian, though I carry a Ukrainian passport.'

One of his grandfathers died at the battle of the Dukla pass in 1944, as a soldier in the Hungarian army, fighting the Russians. Unlike many of the tens of thousands who fell there, he has a grave, with his name on it, in the 'Valley of Death'.

Vasyl worked for thirty years as a tourist guide and retired in 2010. The Hutsul men wore differently coloured hats according to their district, he explains, so you could always tell where they came from. The colour for Rakhiv is green; for the village of Bohdan – brown; and black for Yasinya. The Hutsuls from around Chernivtsi, on the far side of the mountains, have a white band around their hats. Shoes for everyday use were made of pigskin; for special occasions they were fashioned from cow hide. The ornamental axes which men carried on their belts were worn on the left if they were unmarried – 'although they sometimes moved them, to trick the girls,' adds Vasyl.

Unlike the Carpatho-Ruthenians who live only on this side of the mountains – or the Lemkos and Boykos who live on the far side in Poland and Ukraine – the Hutsuls straddle the mountains, with communities in Rakhiv and Yasinya and some seventeen villages on the southwest slopes, but far bigger concentrations on the other side.

'The name Hutsul has taken on a broader and vaguer meaning,' writes Paul Magocsi.

Especially in today's Ukraine it is used as a kind of term of endearment to describe *all* the inhabitants of Ukraine's Transcarpathian oblast [region], who are viewed with nostalgia as pristine mountaineers that ostensibly embody and preserve the best qualities of traditional Ukrainian culture.

He waxes lyrical about a world where little could be found in the shops; where much was made from wood; and where men and women were skilled at making and decorating everything they needed, for their homes and work. When men got married, they were expected to build their furniture themselves: a bed, a table and chairs, and oak barrels to preserve vegetables in, for sheep's cheese to make *bryndza*, or to pickle cabbage. Ash wood was also good for barrels, though it was harder to find. Beech looked good, but fell apart too quickly: beech is best for floorboards, for furniture and for burning. For the handles of the axes the men carry, hornbeam is best.

Vasyl is a little cautious when I ask about Hutsul statehood. 'The best times here were when we were in Czechoslovakia. People were provided with everything they needed, and had a broad choice of what they could do. They could keep cattle, and sheep.' Hutsuls don't always get such a good press, he adds. Some Romanians, for example, deliberately confuse the word *Hutsul* with their own word *hoţ*, meaning 'thief' – and suggest that as an etymology.

He finishes our conversation with a Hutsul story.

Once upon a time, twin brothers fell in love with the same girl. She said she would marry the one who brought her a certain flower. So the brothers set out and started climbing together, up towards Dragobrat, on the edge of Svydovets. On the way, they asked a shepherd where they might find the flower. He indicated a certain rock. The climb was very steep, and in their hurry to get to the top first, both fell to their doom. The next day, the girl came looking for them, and finding them both dead, climbed the rock herself, and leapt to her death. The mountain was so sad, it turned the twins into rocks. And that accounts for the twin-peaked Blyznytsya mountain, which means – 'twins'.

My own initiation into the world of the Hutsuls had come thirty years earlier. Travelling through Poland in the spring of 1993, interviewing people ahead of the election that September, I stayed in the city of Lublin. Walking through the streets one evening, I passed through a gateway to find myself in an alternative theatre. Over beers in the bar, members of the troupe told me that they were planning a field trip to the Hutsul region in the summer, to study Hutsul music and traditions before they were erased by modernity. And so that June I found myself driving east with my friend Roger, in an old dark-red Saab, to meet up with my new Polish friends, who were travelling from Lublin in two minibuses.

Mogur was not at home when we arrived to fetch him from his home in the mountain village of Zelene. He had forgotten that he had agreed to play for us that night at a gathering of musicians, in a dance hall in Verkhovyna. After a while, his wife shuffled up the path wearing a headscarf, suspicious of strangers. 'It's out of the question,' she told my Polish companion, a young violinist from Lublin. Polish is quite close to the Ukrainian vernacular spoken by the Hutsuls, so they could easily understand one another.

'Anyway, he's too old,' she added, without volunteering any information about how old he might be (I later found out he was seventy-three). We patiently made polite conversation with her in her front garden. Zelene is the village where he was born, deep in the forest, 24km from Verkhovyna, but at least an hour's drive.

Then Mogur himself came slowly up the path, carrying two tin pails brimming with apples, and the negotiations started. Mogur's eyes twinkled as we described the scene in the valley below – the musicians who said they would only play if Mogur was there; the crowds of townsfolk, waiting to dance; and the visiting Polish theatre troupe from faraway Lublin, for whom Mogur was a legend. They said his playing was enchanted. They recalled the time he was not invited to play at a wedding because of some imagined slight: in the middle of the wedding party, his violin suddenly appeared out of thin air and started playing. The hired band fell silent, and all the wedding guests gaped.

'Would you play like that for us?' we asked. All around us, birds twittered and the air smelt of fresh apples and mushrooms. His wife was the first to bend. She would allow him to play, just this once, she said, provided she and all the children and grandchildren could come, too. Not a problem! I replied. My Swedish family car might be a five-seater, but the more the merrier!

Fortunately, it was downhill all the way. I lost count of how many adults and small children and babies packed into that car.

The dance hall fell silent when Mogur walked in, wearing his signature grey trilby. The air was so thick, you could have broken off a piece. Then the grey trilby was in his wife's lap, Mogur had whisked his violin from its case, and the room exploded in a cascade of glorious sound.

The dance floor heaved. In the middle of it all, bottles of vodka circled, a single small shot glass on the top of each. It was rude to pour yourself a drink, so a simple ritual had been invented. The person with the bottle moved through the throng, trying to catch the eye of a dancer (which was not hard). The person with the bottle poured a drink, which was downed in one; then the drinker took over the bottle and the glass and started circulating in search of the next taker. A perfect system.

All evening we drank and danced. Musicians joined the band, and left it, to drink or dance. Mogur played throughout.

Eventually it was time to take him and his brood back to the village. Somehow all the children, large and small, were rounded up. Much the worse for wear, I took the wheel. I have no recollection of the journey back up the hill – or down it again, to the small village where we were staying.

Over the following days, the Poles studied local melodies from the violinists, flutists and singers. We ate with local families, mostly mushrooms and potatoes. There was no food in the village shop, only tobacco and vodka. We hiked one afternoon to meet a shaman, who lived beyond the village on the edge of the forest. He could heal illnesses that had been sent by what he called 'the dark forces on earth'. 'What about Chernobyl?' someone asked – it was seven years since the world's worst nuclear accident, in April 1986 in northern Ukraine, near the Belorussian border. Large swathes of Ukraine had been affected by radioactivity, and many people were sick. 'I cannot heal illnesses like that, sent by God,' replied the shaman.

On the way back to Hungary, we picked up a policeman, in uniform, hitchhiking at the side of the road. About an hour down the road, I was stopped for speeding. From the back of the car stepped my ace – my very own policeman. His mysterious and perfectly timed appearance dispelled all danger of having to pay a bribe or a speeding fine.

The master Hutsul woodcarver Yuri Pavlovich has a gallery in the centre of Rakhiv dedicated to his work. There are carvings in astonishing detail, perfectly symmetrical but adorned with faces and scenes, like a storybook. Each curve has a meaning, he explains, as we walk through the gallery. 'If it's rounded like this, it means you're proud; or like this, that you're stubborn. If it curves the other way, you are showing off your power, your strength.' In most of his pictures there are two colours of wood, the white maple and the darker pear.

There are plates and plaques and mirrors, beautifully inlaid. The faces of ancient Hutsul and modern heroes, actors and writers, churchmen and cosmonauts with magnificent Hutsul moustaches. There's even one of the film director Sergei Parajanov, whose film *Shadows of Forgotten Ancestors* is based on a Hutsul love story.

In one picture there's a scene of men loading and tying rafts beside a river. 'This one I carved during my military service, on the Russian steppes,' he explains. He was in the air force. 'I found an old piece of ebony from Africa in a warehouse. No one knew where it came from. But there were no mountains there, no trees, and I was feeling homesick. So I carved this scene.'

Some of his scenes are representations in wood of pictures by the photographer Mykola Sinkovski from the 1920s.

For Hutsul people, everything in nature had a soul. And throughout their lives, they were praying to the good, and fighting the bad in nature. Perun, the thunder god, represented the danger, who might strike you down. And because nature all around them was so beautiful, they tried to make everything they crafted beautiful, so it would fit in. This spirituality of the Hutsuls is being lost in our times, of electronics and space travel . . . I would like my work to act as a reminder of the skills and values of the old times.

Many of those he depicts were chroniclers, writers, scientists — witnesses with the pen or the camera or the knife of an ancient way of life, of a relationship to the land and the weather more akin to a tribe in the rainforests than to the people in a modern city. Some of the faces I recognise. One looks like the man who unlocked a wooden church to show me round, on the hill above Yasinya.

'The main work for Hutsul men was to cut hay for their animals, and wood to heat the home.' He tells a Hutsul legend of how frogs fell from the moon, and brought fire to the people. The capital of the Hutsul lands, on the far side of the mountains, used to be called Zhabie, meaning 'frog', but was renamed Verkhovyna in Soviet times.

The threads in the costumes of men and women worked as a kind of genetic map: 'Boys and girls who liked the look of each other might study the embroidery of each other's clothes, to find out if they were closely related, because that would mean they could not consider marriage.'

There are carvings of eagles and bears, surrounded by magical patterns. The bears were known as 'uncles' – just as in Native American folklore.

Most of the portraits are of men, but there's also one of Ménie Muriel Dowie, a Scottish novelist and travel writer, wearing a headscarf and looking very much like a Hutsul herself. Her journal of her adventures, *A Girl in the Karpathians*, was published in 1892. At the book launch she dressed as a Hutsul peasant girl, and caused a sensation in London.

The metal inlays Yuri uses in many of his pictures are silver wire. 'I bought 10 kilos of it once, from the military airport in Kyiv – enough to last a lifetime!' It cost 36 roubles a kilo.

At the end of the exhibition, the sculptor stands beside an oil portrait of himself, in traditional Hutsul white shirt and trousers, brightly embroidered waistcoat, shoulder bag and ornamental belt, holding a *bartka*, a shepherd's axe with a long handle, with Hoverla looming purple and mysterious over his left shoulder.

'When you describe Hutsul culture, you use the past tense. Is everything lost?' I ask Yuri Pavlovich.

'Not everything,' he replies, thoughtfully. 'At the festivals, you can still see people dressed in their folk costumes. But the thread is not as strong as it used to be. It is no longer made by hand.'

'My colleague rang me from the Academy of Arts. "If you have someone who wants to continue your work," he told me, "I can accept them without any exams!" But there is no one. Young people travel away to the Czech Republic, to Poland or Hungary in search of work.' He himself has no sons or daughters.

Yuri's a figure on the blurred borderline between folklore and folk memory. How to preserve tradition, without turning it into a trinket for tourists, an imitation of itself? It's a dilemma which many ethnic groups around the world have to confront. 'I dedicated my whole life to this work,' he says, in parting. 'I will follow this path to the very end. I am like a dinosaur!'

◎ ◎ ◎

The handsome sky-blue train takes just over four hours to cross the Carpathians from Rakhiv to Kolomyya. For the first part of the journey, it follows the Black Tysa through meadows of wildflowers, past timber-framed houses, the soil black in the ploughed fields. My fellow passengers are mostly locals; but there are some strangers like myself, backpackers from Kyiv, with giant rucksacks. A soldier in camouflage uniform stands to attention holding a rifle at the entrance to a tunnel. A woman in black strides purposefully beside a black Lada car, spectacular against the emerald-coloured trees. The train, pulled by a green locomotive, struggles slowly up the inclines, rumbling and tattling, billowing clouds of diesel fumes. Hayricks are bushy yellow among the green fields, or bare as artists' easels, still to be stacked.

After Yasinya the track switches from the Black Tysa to the Prut, which drains almost all the streams of the eastern Carpathians, almost 1,000km long, to enter the Danube at Galați, just before the Black Sea.

Sitting opposite me on the train is Yevheniy, a retired teacher of literature from Kolomyya, with bright blue eyes, grey hair beneath a flat cap and a handsome face. He wears an old-fashioned striped shirt with big collar, beige trousers and a green jacket. His daughter lives in Italy, he says. It's cold on the train as we go through the mountains, but there's a pile of red blankets to compensate.

Across the carriage, Anastasia, a student of eighteen, wraps herself in a blanket and taps away at a laptop. The train is so slow, I wonder if a branch on the boisterous Prut might get there faster.

Beyond Mykulychyn, the valleys feel more populated, and the train seems to speed up a little. Through Yaremche, with just a glimpse of its famous waterfall, where the Prut falls 5 metres. White ducklings scatter at our approach. We pass a church and Yevheniy crosses himself. Anastasia puts aside her laptop and starts weaving a green thread round a wickerwork ring, looped around with white string. A squall of rain drums on the windows of the train. A man with a yellow flag stands, getting soaked at a level crossing, beside a bush thick with dog roses.

Rostiv is waiting for me on the platform at Kolomyya. He's a Hutsul from Kosiv, 'though people call us *Boyki* – the diminutive of *Boyko* – if they want to be unkind'. An example of the blurring of Lemko, Boyko, Hutsul and Ruthene, in a world where none is clearly defined.

The local authorities, Rostiv explains, are keen on 'development'. We're still in the middle of the Covid pandemic (summer 2021), but the guests keep coming. As a guesthouse owner, he can't complain. 'We follow government instructions – lots of cleaning fluid on the tables.'

We eat at the Arkan restaurant in Sheshory, on the shore of the Pistyn'ka river. 'Traditional Hutsul cuisine with a modern twist,' reads the blurb, and we're not disappointed.

Kosiv once had a large Jewish population, and was a centre of the Hasidim, the followers of a mystical form of Judaism. The rabbi Baal Shem Tov came here in the eighteenth century. His name means 'a good man who knows the secret name of God'. There was a popular song:

Between Kosiv and Kuty, there is a bridge,
Where Baal Shem used to stroll . . .

He wandered through eastern Galicia, worked as a labourer in the clay mines, married twice and meditated. His reputation grew as a healer and miracle worker, and often drew the wrath of the authorities. He is said to have met the Hutsul outlaw Oleksa Dovbush and to have received the gift of a pipe from him, in gratitude for hiding him from the Polish authorities.

Once, the Baal Shem Tov asked Dovbush, 'How much longer do you plan to be a robber? You have so much goodness and love for all creation in you; can't you settle down somewhere, work the land, build a house, take a wife, have children, raise them and just live, as God commanded? Why do you sin so much?'

'Recruiters and commissioners came to conscript us,' Dovbush replies. 'They cut off our long curly hair, put us in tight uniforms, and took us far from home, to Vienna, beyond the deep Danube, where even ravens would not find our bones. They brought us to Kolomyya, where we were guarded by soldiers. We talked long into the night, girded our loins, and then we attacked the guards, tied them up, and ran away back to our mountains with their weapons. From then on, we have been fighting them and their laws. The fight is hard, and our life is bitter.'

Mykola Strynadiuk has a lot in common with his fellow Hutsul woodcarver Yuri Pavlovich on the far side of the mountains, but his style is very different. There's an irreverent, modern humour in his work. He colours the wood in his carvings, and incorporates bits of everyday items, eggshells or pieces of leather into his work. Unlike the Rakhiv master, his pictures are often asymmetrical: a triptych of three sleeping shepherds under three sickle moons, who have dozed off while watching over their flocks at night, their brown hats pulled down over their eyes. He carves on linden blocks, rather than maple and pear.

One picture looks like a traditional Hutsul carved cross. Closer inspection reveals an amazing mosaic of broken eggshells from painted

Easter eggs. Spruce trees rear up against dark mountains, with a sky full of stars and a huge moon coming up over a triple-peaked mountain. Hutsul hats are everywhere, all black, as if from Yasinya, but with different coloured and patterned bands round the middle, pulled down over their wearers' eyes.

'Would an old Hutsul laugh, cry or punch you if he saw this picture?' I ask.

'I'm pretty sure he would laugh.' A man stands with two cherries dangling from his teeth, and a white shirt tied with some very playful yellow and orange tassels, like circus tents. 'The sexy Hutsul, I call it. Tradition can sometimes be conservative, can act as a sort of censor over one's art. I like to create my own traditions by myself.'

One carving depicts a typical mountain landscape, spruce trees against the yellow-brown mountains. In the foreground, a pale-green mountain pasture is peppered with wildflowers, made of coloured circles of wool.

Finally, he shows me his collection of old wooden icons. They are on simple boards, naive depictions of the saints, especially Saint Nicholas, who is especially venerated by the Hutsuls. Mary, Jesus and St Nicholas all manage to look rather Hutsul-like, their golden haloes set off nicely against a dark-green background, dotted with mountain flowers.

We sit down to a big spread prepared by Mykola's wife: bread spread with caviar, cakes, ice cream, tea and *horilka* – the Ukrainian equivalent of the Hungarian *pálinka* fruit brandy, flavoured with herbs and spices.

◉ ◉ ◉

On my visit in 2021, I seek traces of Mogur, and discover that he passed away in 1997, four years after I met him. Maria Sonevytsky's Columbia University dissertation, 'Wild Music: Ideologies of Exoticism in Two Ukrainian Borderlands', is the best source.

She contrasts the folk traditions among musicians in the Crimea and the Hutsul lands – on two peripheries of Ukrainian culture:

Both of these groups . . . embody dominant stereotypes of otherness in Ukraine – Hutsuls as the ideal Herderian romantic folk, and Crimean Tatars as the menacing, mysterious, 'oriental' other.

She spoke to several Hutsul musicians who knew Mogur, including his 'best pupil', known as Hurduz. The following details are taken from her research.

Mogur's real name was Vasyl Ivanovych Hrymaliuk. Born in March 1920 in Zelene, his mother opposed him learning the violin as a child, because 'musicians never make good men', and smashed the instrument he had saved up for. He ran away from home and was brought up by Dmytro and Kateryna Mohoruk, from whom he adopted his musician's name 'Mogur'. After hearing him play, a Polish army officer gave him a violin which had belonged to his own dead child. He made his gift conditional on one thing – that Mogur should never lie.

This made a deep impression on the young man. He studied under the renowned violinist Gavitz in the village of Iltsia. As his reputation as a musician grew, he became much in demand at weddings and baptisms. Stories of the magical power of his music began to spread through the valleys.

The art of making music, Mogur explained in 1995 to Roman Kumlyk, a violinist from Verkhovyna, and local historian Ivan Zelenchuk, is a gift from God: 'Music has a great power; it helps people to live, and to become better.'

According to Sonevytsky, Mogur's first true love, about whom he spoke for the rest of his life, was Yidvokha Filypchuchina, who played the *drymba*, or Jew's harp. She was murdered by the Red Army when they swept through the Hutsul region, setting fire to villages in the late summer of 1944.

Mogur composed a melody for her which saved his life. As a professional musician after the Soviet occupation, he had to play for the new masters. On one occasion, he was walking near his village when he heard partisans from the Ukrainian Insurgent Army (UPA) coming down the forest track. He tried to hide, but they spotted him. Recognising him as a musician who had played at Communist Party events, they decided to execute him on the spot. He was given one last wish. To play for them, he said. So he took out his violin, and played his melody for Yidvokha. He played so beautifully that the partisans wept and spared his life.

Maria Sonevytsky tracked down his pupil Hurduz, real name Vasyl Iliuk. His wife Maria told the following story. When she was pregnant with her first son, Vasyl, Mogur arrived to pick up Hurduz to go to a concert together. Mogur turned to her and said she faced a difficult labour at home, so she should go immediately to the nearest hospital. Hurduz took her there, against her will, as she felt fine. The labour began, and as Mogur had predicted, was long and hard. The baby was saved.

'He read it in the cards. He could help with livestock, if the goats and cows weren't giving milk. People in the community would consult with him.'

Many of the legends about Mogur feature his rivalry with other musicians and his supposedly magical powers: when he played at competitions between musicians at weddings or in 'cutting contests' (*perehrakh*), strings would sometimes break – a bad omen. And sometimes musicians who played the *bayan*, a kind of accordion, with Mogur, lost consciousness and their instruments would break.

When the film director Sergei Parajanov came to the region in 1964 to make *Shadows of Forgotten Ancestors*, he chose Mogur as his musical director. The plot is based on a novel by Mykhailo Kotsiubynsky. The young Ivan rushes through the snowy forest to bring his older brother lunch. The tree his brother is cutting begins to fall as he approaches, and the older brother is crushed, saving his young brother from the falling tree. Ivan falls in love with Marichka, the daughter of a family who are bitter enemies of his own family.

Sixty years on, the film is fondly remembered in the region, and a house in Verkhovyna is dedicated to memorabilia and photos from the shoot.

Vasyl Kischuk was born in 1935. His home clings to the hillside, the front terrace facing south, where he sits on a bench, repairing a *trembita*, the Hutsul trumpet. The wood should be a strong pine branch, he explains, the only tree long and straight enough for the job. He takes a piece, cuts it to the right length, then splits it exactly in half – a delicate task. He hollows out the two halves with curved chisels. Then he binds the two tightly together with birch bark, soaked in glue made from boiling bones.

He has dressed up in his best embroidered white smock for our visit, patterned with red and brown and white symmetrical stars down the front. His ears stick out beneath his smart brown hat. His face is less lined than his eighty-six years might have led one to expect, but his hands are scarred from the work with sharp knives. His workshop is a magician's cavern of axes, scythes, chisels and hammers, clamps and saws. The base of each chisel is worn away by decades of hammering. There's a curved hunter's horn, wrapped in birch bark, into which he inserts a neat mouthpiece, worn on a chain over the shoulder, ready to be brought swiftly to the lips even on horseback. There's an old saucepan in which he boils cow bones for the glue. From the doorway of the kitchen, his wife Maria and one of his grown-up daughters – both in traditional dress – listen to us. The house they live in was built by Maria's father.

Three or four *trembitas* are sounded at a funeral, Vasyl explains. They are mostly used at weddings and for Christmas carolling. The lads of the village gather outside the church at eight in the morning, after the early mass, then go from house to house, summoning the baby Jesus from the depths of the winter snow.

The English-language website of the *Ukraïner* periodical carries a beautiful description of the ceremony, known as *koliada* – a mid-winter festival to celebrate the birth of the sun:

On Sunday morning, an early sunrise
Brought joy and glory to cherry gardens.

The cherry blossom tender and white,
Oh, little Jesus was born tonight.

The carolling starts with a short ritual dance, the *plyes*. Then the
company troops from house to house, behind a leader known as a
bereza, and are welcomed in, given food and drink, and invited to play.
Usually they receive money for the church. In Soviet times, the practice
was banned, but continued quietly in high mountain villages, though it
is hard to imagine a quiet *koliada*.

There's only one dance with the house owner. It's called *kruhliek*. It's
danced in a circle, and if the house owner has bee hives, they carol a
special song for the bees. When the carolers come to an abandoned
house where nobody lives, they will still carol there, so that the life
comes back to that house . . .

Among the carols . . . [one] has the following words:

The rooster is cut and salted
And the fish is boiled and salted,
And this rooster whose head was cut off,
And this fish, cut and salted, and boiled.
When the rooster crows without his head,
And this cut, boiled and salted fish will jump in the water and
     swim away,
That's when Jesus Christ will resurrect.

The deep voice of the *trembita* signalled the main feast days of the
year, especially St George's day, to announce the victory of spring over
winter. The shepherds went from house to house, collecting the animals
from their owners, to lead them up onto the high mountain pastures
for the summer.

Out on his terrace, Vasyl gives a demonstration, and a long, deep,
mournful sound fills the meadow. In *Shadows of Forgotten Ancestors*, the

sound of several *trembitas* is a frequent refrain, fuelling the tension and drama of the scenes. At the end, at the wake for the hero Ivan, *horilka* is poured into the end of the *trembita* 'to warm its throat' – as before the *koliada.*

Vasyl hands me the instrument. Only upland shepherds can make it sing, I have heard, so I left it to my lips with trepidation. It is surprisingly light, for something that is 3 metres long. I purse my lips and blow hard. Nothing but a feeble, fart-like sound. 'You have to sort of hum into it, not just blow,' explains Vasyl. I try again. To my delight and relief, a respectable horn-like blast seems to build up inside, then pour out of the trumpet.

Then we sit down in the kitchen. The table groans under the weight of another Hutsul feast. Hard-boiled eggs, a great bowl of soft, creamy *bryndza*, and another of more mature, crumbly cheese, tomatoes and cucumbers, slices of salty and sweet breads, a maize-like cake spread with plum jam, and various meats. In the middle of it all, a bottle of home-made *horilka* with a deliciously bitter taste. When I ask what is in it, Vasyl disappears outside, then reappears at the table with a great bunch of stinging nettles in his big chiselled hands. It's his party trick – the leaves no longer hurt him. Instead of stinging my mouth, the drink begins to sting my mind, in the most pleasant way.

Vasyl's mother died when he was eight, during the war; his father when he was twelve. Everything he knows about making musical instruments he taught himself. He has three goats and two horses.

The room is simply decorated. A bright-blue bed with big cushions, matching wooden chairs and stools, flower-rimmed plates mounted on the wall above delicate embroideries, a big wood-fired kitchen range bedecked with pots, a blue chimney reaching up through the wooden ceiling. Vasyl's wife watches over us all like a little bird, following every movement of fork or spoon or cup to lips, making sure her guests are properly fed. There are fading photographs on the wall of men in uniform; chubby-cheeked girls with long plaits; a solid, unsmiling couple, the moustache at the centre of the man's face, like an aeroplane propeller, and the woman scowling a little at the intrusion of alien technology into her routines; another couple – a young man with big ears in traditional leather waistcoat, beside a knowing, headscarved woman in yellow blouse and green cardigan. There is also a good collection of babies and icons – the Virgin and Child, a scene from the Last Supper and a portrait of Christ Pantocrator, with a quotation from John's Gospel, Chapter 8, verse 12: 'I am the light of the world . . .'

Vasyl still remembers Mogur. 'He could look in various places at the same time,' he says; it's not clear whether he is referring to his magical powers or to the fact that he was a little cross-eyed.

Mykhailo Tafiychuk is eighty-five, another near contemporary of Mogur, in Velyki Bukovets. He played against him once in a violin competition – 'and Mogur ran away!' There's a lot of rivalry between the musicians, up here in the hills. The main difference is that Mykhailo also makes violins, while Mogur just played them, he says.

We start in Mykhailo's workshop, another den of home-made chisels, saws and shepherd's flutes. He loves his work, and can never sit still. He recently worked the whole night through, finishing an instrument. He picks up a long flute, a *fujara*, similar to the ones Samo Hríbik makes, and starts playing. On the recording I made of him there are lambs to be heard bleating in the background and a dog barking.

He learnt most about making instruments from Mohuruk, who died at the age of 100. Another strong influence was a blind player of what he calls the *lira* – the 'wheel lyre', or hurdy-gurdy in English – whom he heard playing and singing when he was twelve. Inside his living room, he takes out a lyre, and sings a song he learnt from the blind musician, seventy years earlier. 'It's a song about a conversation between Jesus and a man who has sinned, whose sins cannot be forgiven.'

'It's an ideal instrument if you only have one arm,' says Mykhailo. It sounds like something between a diesel engine and an orchestra.

He went to New York once with a bag of flutes he had made, 'and I didn't come back with a single one. They wanted me to stay there, and in Germany, but I wanted to come home.'

He refuses to use electrically powered machines, and works only with hand tools. Some of the instruments he makes, like the *kobza*, are from a single piece of wood, hollowed out.

'The most important thing is to love the work you do,' he says. 'If you really want to do something, then you can. Your desire is what matters.' His sons and grandchildren all play instruments, but none of them wants to learn how to make them. That saddens him, but it can't be helped. He has no idea who will make musical instruments in the future, but is not too concerned: he's done his bit. Inside the house, beside the dark-brown tiled stove where he warms his freezing hands, he pumps up a *duda*, an eastern European bagpipe.

He finds teaching difficult, because the young musicians who come to him want to play their own way, and he wants to teach them his way.

'And what is your way?' I ask, cautiously.

'Music comes from God,' he says. He's not sure even if it can be taught. Certainly not to people who are not as immersed in it as he is. He plays all day long, whatever instrument is to hand. In the two hours we are together, he plays all manner of flutes, the bagpipes, the wheel lyre and the violin. The violin, his pride and joy, hurts him now. 'But I can still make them!' he says, defiantly.

Mykhailo doesn't drink or smoke, and has been married to his wife for sixty years. Before we leave, he plays a love song on a small flute, a *sopilka*. I look back across a meadow of wildflowers to wave to him, silhouetted in the doorway.

⊙ ⊙ ⊙

On my last day in *Hutsulschyna*, the land of the Hutsuls, I hike with friends up into the hills above Yavoriv. We cross a wooden bridge over a stream, past a trout farm and then up through a wood to the *polonina*, the high uplands, emerald meadows brimming with purple irises and other wildflowers. The Chornohora ridge, crowned by Hoverla, rises up to the west. It is a clear day with a brilliant blue sky. The purpose of the trip is to visit Vasyl, a shepherd, to buy cheese. The sun is still out when we reach the three buildings on a gently sloping ridge where he lives and keeps his livestock – forty sheep and two cows of his own, plus eight more cows that he looks after for others.

Vasyl is sixty-two. We stand around in his cavernous shed, admiring his cheeses on a long wooden shelf. They're all rounded, in lighter or darker shades of yellow, and have a pleasing, slightly irregular surface, like the craters of the moon. It's quite dark and chill in the room, the only light coming through the open doorway at the end. As my eyes get used to the gloom, I see more details of the room, the wooden pails and the cheeses. The girls huddle deeper into their coats. Vasyl wears a striped woolly hat, summer and winter, and speaks in the loud voice of someone who spends most of his life outside, bellowing over the wind. We hear the first patter of the rain.

As a child, Vasyl used to walk 8km a day to the primary school in Yavoriv. Now he milks the animals twice a day, morning and evening, and makes cheese each morning: four parts of cows' milk to one of sheeps'. He uses the rennet from a calf's stomach to make the cheese set. He also makes butter and *vurda* (*urdă* in Romanian, *orda* in Hungarian) – the product that results from boiling up the whey left over from making cheese. It comes in solid, crumbling blocks – a little dull on its own, but perfect in pancakes and pies. He sells his cheese to tourists like us, and once a week takes them down to the market in Kosiv, where they fetch $3 a kilo. His daughters are both married and live in Poland, though he can't remember quite where. They come back whenever they get a visa. He went abroad himself once – by mistake, he laughs. He wandered with his sheep over into Romania, without a passport. He has no desire to travel: he's happy where he is, with the harsh weather and the steady, predictable routines. He's aware of climate change not so much in the weather, as in the steady march of the trees up the mountainsides – oak and beech and spruce pushing the treeline ever higher. Another change in his lifetime has been the erosion of the mountain pastures by shrubs, in the absence of animals to graze them.

His daughters would only come back to Ukraine if there was a significant increase in salaries. He doesn't blame the young for not wanting to copy his lifestyle. This is what he has always done, 'because it's all I know'. If no one takes over his farm, it will slowly crumble into the dust, he says, without emotion.

On the way back down to the village, a sudden squall of rain catches us out and we get thoroughly soaked. The thunder rolls and the lightning crackles. When the sun emerges briefly, I take a picture of the girls, laughing beside a bush of purple dog roses.

CHAPTER 10

THE WORLD STAINED RED

My head is rolling from grove to grove
like tumbleweed
or a ball
my torn-off arms
will sprout violets in the spring
my legs
will be pulled apart by dogs and cats
my blood
will stain the world a new red.

Maksym Kryvtsov (1990–2024), Ukrainian poet

Eight months after my first visit to Transcarpathia, I wake in my Budapest flat at 04.30 to discover that the Russian military has just launched a full-scale invasion of Ukraine. I pack an overnight bag, camera, tripod and radio equipment and wake up my teenage son, Jack. We drive down to the border, at Beregsurány, to see the refugee flows. And there they are, mostly young men, trudging down the road into Hungary, pulling their ungainly suitcases on small wheels, as if on their way to Budapest airport for a cheap flight to London for the weekend.

While hundreds of thousands of Ukrainians engage in a desperate attempt to defend their homeland, tens of thousands flee across the borders into Poland, Slovakia, Hungary, Romania and Moldova. Who in the twenty-first century is willing to risk their lives to defend their country? A survey in 2015 suggested that 62 per cent of Ukrainians would, compared to 74 per cent of Finns, 27 per cent of Britons and only 18 per cent of Germans. Moroccans led the list, at 94 per cent.

That night, Ukrainian border guards prevent all young men between the ages of eighteen and sixty-five from leaving, unless they have a serious medical condition or at least three children.

For the next month, I travel the Hungary–Ukraine and Romania–Ukraine border, talking to the refugees. After the first day, they are mostly women and children. Many are from Transcarpathia. As the westernmost region of the country, with perhaps the weakest sense of Ukrainian nationhood, they have too little in common with Ukraine, even in its hour of need.

Taras, my friend from the ridge walk, went the other way – straight to the eastern front as a volunteer, even though he had no military training. 'On the first day of a full-scale war, I went to the army,' he wrote on Facebook in January 2023. 'Exactly a year later I retired from the Armed Forces. Ten incredible months in a combat zone alongside the same crazy volunteers. Without any rotation, almost all the time at "zero".' Zero is slang for 'ground zero' – the front line.

In February 2023 I travelled back to Transcarpathia for the first time since the invasion. This has been the region least affected by the war, but it has absorbed hundreds of thousands of Ukrainians from the rest of the country. These are people who chose to sit out the conflict in the safest part of the country.

As my number 6 tram crosses the Petőfi bridge over the Danube in Budapest at dawn, the clouds to the east are wild and dishevelled, the shape of a galloping wild boar. My carriage at the West Station in Budapest is half full, mostly of Ukrainian women. There are also a few men milling around, saying goodbye to the women. If they went back to Ukraine, they might be conscripted. Some of the women are laughing, others crying; but most are silent. One young woman is distraught when her man gets off the train. She goes to the window. When he finally walks off, she takes out her phone and starts a video call with him straight away. How little space we allow ourselves nowadays to be alone.

Several women are travelling back with children, who gaze around or play on their phones. The younger ones are wide-eyed. Going back to their homeland is an adventure for them, part of the inexplicable grown-up world. Many of the women have cats or small dogs in baskets.

The Latorca intercity leaves exactly on time, at 07.24, as though to reassure the passengers that even in wartime some aspects of life are reliable and predictable. The track is like a thread, drawn tight across Europe, and the train a needle, embroidering patterns on the fabric of the continent. We took peace and security for granted for so long. Now the fabric is ruffled, and everything feels threatened.

The sun bursts in long shafts through the train, sometimes from one side, sometimes the other. The first sunshine for weeks lightens the mood of the passengers.

The sandy soil and vast skies of the Hungarian plain give off their own mysteries; but this morning they're just a stepping stone to the war. Even the Hungarian ticket collector seems kinder to the passengers than usual, as though he feels sorry for them. How much solidarity there is between human beings, when politics doesn't turn them against each other.

A Roma woman with big earrings and wearing what looks like a fluorescent pink dressing-gown strikes up a conversation. She's from Mukachevo, and cannot connect her phone to the internet to speak to her son, who is waiting for her at the other end. Another passenger creates a hotspot, and she makes contact. The confusion concerns whether the Ukrainian arrival time on the Hungarian timetable is Kyiv time (one hour ahead) or Budapest time. Much of the population of Transcarpathia calculate their lives in Central European Time, in order to distance themselves from the events in the rest of the country.

The Roma woman has been in Hungary for a year, has had an operation to remove a malignant tumour and is getting chemotherapy at the hospital in Kecskemét. She's actually not well enough to travel, but hasn't seen her sons for so long, she has to go. They cannot leave the country, of course. She's nervous about the border, she confides, because her papers are not completely in order.

Four hours later, at Záhony, her fears are confirmed. Two young border policemen flick through her blue Ukrainian passport, pull faces, then tell her she has to get off the train. She pretends to be surprised, then indignant; she gives them an earful about her medical problems, but reluctantly assembles her many bags – an ungainly suitcase, a nylon sack and various bulging plastic bags, tied with string. I get up to help her, awkwardly aware that I am helping facilitate an eviction, but the policemen are not going to help her. Fortunately for her, the train stands at Záhony for nearly an hour, as the locomotive is changed.

After forty-five minutes she's back, triumphant: 'The policemen on the train were monsters, but those inside the station were angels! "Am I banned?" I asked them. "No, darling, you can come back whenever you want!" the officer told me. I could have hugged him!' The train, reduced to just two stubby carriages, leaves almost immediately.

We slide over the Tisza rail bridge, the river a military green beneath us. A Chinese man, one of the few males on the train, catches my eye for the first time, and nods. We all have our different, private reasons for being glad to be back.

The passengers are requested to get off at Chop, and join a long queue to passport control and customs. We shuffle steadily forwards. I appear to be the only person without a dog or cat. The queue moves swiftly, and after only ten minutes I reach the front. A sharp-eyed border guard studies my dark-blue British passport. Tourist? He asks. 'Yes, but with a purpose . . .' I say. 'Visiting friends . . .'

I'm not here as a journalist, so have not applied for journalistic accreditation. 'Have you visited our country before?' 'Yes . . .' and I offer a date, but no more details. 'Welcome to Ukraine!'

A young woman in army fatigues asks me to put my rucksack on a table, and open it. Alcohol? She asks. I confess to one bottle of red wine and half a litre of home-made *pálinka*, both gifts. 'That's normal,' she grins, and waves me through. I'm back.

Outside in the station forecourt, I brush aside taxi drivers and squeeze onto the little bus to Uzhhorod. There's nowhere to change money, so I offer the driver a grubby €5 note. He is unimpressed. No one on the bus wants to change it into hryvnia. Then the bus driver passes it briskly through the window to another man, and suddenly I'm the proud owner of both a bus ticket and 140 hryvnia in change. The bus, packed to the gunnels, swerves out among the potholes of the city.

Chop, as a major road and rail junction, was the site of ugly tank and artillery battles at the end of the Second World War. It still feels wounded today, bringing to mind all the new Chops to the east, senselessly blown apart in the latest conflict.

I get off at the main bus station in Uzhhorod, and take an ancient taxi to the centre. The driver is an Armenian who married a local girl and has been here for decades. At a set of traffic lights, he reaches up and takes his yellow cab sign off the roof. When we arrive, he cautions me to hide my wallet until a police patrol car has passed. The journey costs 100 hryvnia, all of €2. I settle into a comfortable café near the footbridge over the Uzh, and ring Oleksandr.

Ukraine has changed a lot since I was last here. Now it's a country mobilised for war. Oleksandr's bicycle shop still has two workers, but people have no money to buy bikes. If they have any savings, they buy a generator, to get them through the constant power cuts. He can keep his staff on because they do repairs. His flat is on the top floor of a four-storey block, out near the airport. Gabriela and Solomiya are at home. Their youngest daughter, Olivia, will come back after six.

They show me the latest electricity chart on their phone. There are three colours: green, for the hours when there will be electricity as usual; red for the hours when there will be none; and grey for the hours when there might, or might not be.

It's been a winter of constant Russian missile attacks on the Ukrainian energy infrastructure. The Russians are targeting the power distribution network, in an attempt to wear down the population's will to resist. Much of Ukraine's electricity comes from its five nuclear power stations, the largest of which, Zaporizhzhia, is in Russian hands. In the rest of the country, electricity is rationed, district by district, block by block. Oleksandr fills two thermos flasks with boiling water ahead of the evening power cut, and the family charge their phones.

Another way the war impacts Uzhhorod is the air-raid sirens, sometimes four a day. The Russian missiles are launched from far away. No one in Ukraine knows where they are targeted or if they will change course in mid-air, and so the sirens ring out all over Ukraine.

Snowflakes start falling as we walk through the cobbled streets. On a railing, a tiny Father Christmas statue, made by Mykhailo Kolodko, hauls a sack of gifts. Each year a letter box appears here by magic, and children deliver their requests to Father Christmas. Christmas has been another piece of collateral damage in the war. Transcarpathians usually celebrate Christmas twice: the Latin rite on 24 December and the Orthodox on 6 January. As the schism between the Ukrainian and Russian Orthodox churches widens, and as the Ukrainian population turns against everything with even a lingering flavour of Russia, the 6 January Christmas is rapidly losing ground.

The former synagogue glows red beside the dark-grey river in a damp, snowy dusk. Beside it, a six-pointed Star of David is set into the pavement, planted with six spindly weeping willows. There is also a tall, iron column composed of layers, each representing a local family lost in the Holocaust.

The castle is closed for the day, and two armed security guards loiter on the drawbridge. Oleksandr points out a mural from the mid-twentieth century: four figures, wielding tools, with a pile of sacks: 'As children, in the last years of communism, we thought these figures were proud peasants, their sacks full of potatoes. Only after independence in 1991 did we realise that these were the famous brigands, Oleksa Dovbush and his companions, and that the sacks contain treasure.'

Outside the neat brick building of the arts school (and the unlikely-looking centre for nuclear research), there's a beautiful mosaic of a girl on a low brick wall, partly hidden by leaves and branches – a student who helped make the mural. Oleksandr shines a torch on her. The puzzle of her proud face glows amidst the undergrowth. The 1960s seem as far away as ancient Troy.

We settle into a wine bar called Siro Vina. The name is a play on words: read separately, the words suggest 'cheese' and 'wine'; but as a single word, *sirovina* is 'raw stuff'.

We taste a dry Furmint from the Muska winery near Mukachevo, with a full-bodied, spicy flavour. There are fresh tulips on the table and mellow piano music. The glasses are elegant, and the whole place radiates the taste and prosperity of pre-war Uzhhorod. But before which war?

Then a red from a local winery. It's a bit sour at first, but improves as the glass warms in the palm. I'm struck by the strange juxtaposition of war and peace.

Oleksandr is willing to fight, and has many friends on the front. In the year since the Russian invasion, he has passed two medical examinations, and is now on a list to be called up at any time.

'This is not how I imagined my work in the twenty-first century,' he says quietly. 'But many of those on the front have children, too.' People with businesses support those fighting with donations of money or equipment, food packages, warm winter clothing and drones. The roads east through Ukraine are filled with west Ukrainians doing delivery runs to their comrades on the front line.

He shows me the profile of a Ukrainian sniper he follows, nick-named 'Robert Magyar' (*Мадяр*), from Uzhhorod. His haystack of hair turns out, on closer inspection, to be camouflage netting. Each post is written in Ukrainian, and some are also in Hungarian or English. He has almost a quarter of a million followers. In his latest post, he lambasts

Hungarian Prime Minister Viktor Orbán for describing Ukraine as 'a no man's land', comparable to Afghanistan.

'Russia's goal is to make Ukraine an ungovernable wreck, so the West cannot claim it as a prize. At this, they have already succeeded,' Orbán told a gathering of conservative journalists in Budapest. 'The West doesn't understand that time is on Russia's side in Ukraine. Russia is a huge country, and can mobilise a vast army. Ukraine is already running out of troops. When that happens, then what?'

'Come here to Bakhmut,' responds 'Magyar', 'and tell the Hungarians of Transcarpathia, fighting here now, how you and Putin plan to divide Ukraine.'

'Magyar' posts a new video every day on Telegram, and his followers pledge money to buy kamikaze drones. Some $25,000 was raised in just an hour and a half – enough for eight of them.

Oleksandr's parents both came from Khmelnytskyi, 500km further east, which still counts as 'western Ukraine' in this vast country. His father was an engineer, his mother a biologist. They met as students at university in Uzhhorod, lived in Kharkiv for a while, then moved back.

We wander through streets plunged into darkness by power cuts, where only the car headlights and the torches of pedestrians, or occasionally a light in the window of a shop that boasts a generator, relieve the gloom.

It's like the first months of the war in Croatia, in the winter of 1991. When the train I was on from Budapest crossed the border from Hungary into Croatia, all the lights were turned off and we slid through villages and towns, and into the capital Zagreb, in total darkness. It felt like a ghost train entering a ghost country. Only the sounds, and the hard form of the carriage, seemed real.

At home in Uzhhorod, Olivia is back from kindergarten and has snuggled into bed with her mother and sister to watch cartoons on television in the last hour before the lights go off.

◎ ◎ ◎

The road climbs a shallow incline, up into the hills towards Mukachevo, through forests grey and leafless in the snow. There's a white dog in an enclosure by the vineyard, barking fiercely. Vlad, in red winter jacket, walks up the hill towards us. He's pruning three hectares on his own with electric secateurs attached to a battery on his back. He shows me his technique – the Guyot method – leaving just the main stem on each vine. He leaves the cut branches where they are, for his mother to collect later.

As he cuts, we talk. When he bought the land, he was a project manager with a cross-border NGO, weary of office work, with no previous experience of wine-making. His land was completely overgrown by the forest.

Making wine has been harder than he expected. He started with a €300,000 grant from the EU, and planted three grape varieties best known in the Tokaj region of neighbouring Hungary – Furmint, Hárslevelű and Sárgamuskotály (Yellow Muscat). The climate is similar to Tokaj, where the Bodrog and Hernád rivers generate a mist that hovers over the valley in October and November, to shrivel the grapes into currants and infect them with 'noble rot', which gives the wine its special taste. Vlad has the Latorica river, flowing along the valley, and the woods to keep the air damp. The humidity here is even higher than in Tokaj, he explains – up to 70 per cent, whereas in Tokaj it hovers at around 40 per cent. That means that in a rainy autumn, there is a danger that all his grapes will rot, rather than just those whose job it is to infuse the others.

The snow is still beautiful between the vines. 'When I walk here, I see all the blemishes, all the mistakes. But when I fly a drone, everything looks perfect from a certain height.'

He's not allowed to fly his drone anymore, because of the war. He has other problems, too. The global market has convinced the whole world that it wants dry wines. People come here and ask for a dry red. 'Dry red, dry red, dry red . . .,' he parrots. 'When I tell them about my Cabernet Sauvignon – a semi-dry – they refuse to even taste it, and drive off!'

As the metal door of his cellar groans open, a marvellous aroma of 10,000 litres of wine washes over us. The cellar is like a shrine, lined with 620-litre stainless-steel tanks and a row of wooden barrels along the back, made from Hungarian oak.

We start with the driest whites, in this sweet kingdom: the light grape known in Hungary as Hárslevelű, and by the Ukrainians as Lipovene – the linden of the Slavs. The taste is green and refreshing, but with a hint of sweetness on the tongue.

The others chatter in Ukrainian. I catch a word here and there, 'White Burgundy', 'Rheinhessen', but mostly the Ukrainian washes over me. There is an intimacy in not understanding a language and listening instead to the cadence of the words, the sounds and the scent of the cellar, and the gurgling of the pipes.

'In this one, the semi-dry,' says Vlad, 'the difference in sugar levels is only half a gram. Three grams per litre in the dry, three and a half in the semi-dry.'

The Sárgamuskotály is next. A deep, golden yellow which sinks back all the way through the mouth and curls up into the head to infuse the skull.

I notice he is drinking something a little cloudier. 'I was bottling wine this morning. This is the leftover,' he grins. 'Too good to pour away.'

'Was 2022 a good year?' I ask, stupidly.

A shadow creeps over the faces of my Ukrainian friends. How could a year in which their country was invaded, in which tens of thousands of people were killed unnecessarily, be a good year?

Next comes the Furmint, a semi-sweet – Vlad's favourite. 'But people don't like it . . . I don't know why. It's acid and sweet at the same time. Acidity is 8, which is not too much. It's sweet and you can taste this noble rot aroma.'

Next, a Lipovene after three years in a Hungarian oak barrel. Hungarian barrels are even better than the French, he says. The wood is seasoned for three years, then 'toasted' in three stages. Then it is soaked and bent into shape. They also make oak barrels here in Transcarpathia, but the wood is raw, and the woody taste overcomes the wine.

Next comes the Cabernet Sauvignon rosé. If we were bathed in sunlight, above the ground, I would hold the glass up to the light. Down here we are blind to subtleties of orange and pink, and focus instead on the taste. Above ground everything would be different. Glasses would be poured and clinked to victory, and to loved ones fighting, or lost in battle. Instead, the wine cellar feels like a bomb shelter. We just taste, and listen. The glasses are raised, but never kiss.

Finally the Cabernet Sauvignon. Flavours of blackcurrant, rhubarb, raspberry. We emerge laughing from the cellar, into the crisp white air and the white snow and the white dog barking. There's a single, long-haired cat curled up on the wooden beam – also white.

'The war has been a disaster,' says Vlad. 'It ruined my personal life completely. I don't have one!' After two years of the Covid pandemic, he dared to dream that business would flourish. Instead, they're down to just 10 per cent of pre-war levels.

'I tried making juice.' He waves at a wall of juices, arranged according to grapes of single origin. His wines cost only €3 or €4 a bottle, and his juices even less; but he can't sell them. 'People don't want to buy because they don't know what will happen tomorrow. They might need the money, even the small money, for something else.'

Another issue is conscription. He's forty-two, and could be called up at any time. But he has no one who can prune the vines, or make the wine in his place. His mum helps, but that is it.

'The war is destroying our lives. And at the same time, we see corruption all over Ukraine.' That week there were several arrests and dismissals from public life over corruption allegations. For a while, people avoided discussing corruption, as criticism of one's own side in wartime sounds like enemy propaganda.

'What makes me mad is that they are gathering people off the streets to take to the war, yet at the same time others are getting rich selling humanitarian aid. If they want us to fight for our country, they should create a strong one, a patriotic one.'

We drive back to Uzhhorod on pot-holed roads, past giant billboards of soldiers who have died in combat, alongside advertisements for private hospitals. Two worlds, mixed up together.

◎ ◎ ◎

Oreste's friend Serhii fetches me from the bus station in Khust. The slush of the valley turns white and crisp as we climb the shallow hill to Nyzhnje Selyshche. We swerve to dodge the vehicles racing down the narrow road where the Gypsy children sold wild strawberries when I was last here.

The kitchen at the farm is full of people – Oreste's wife Jolana, a Berber from the borderlands between Algeria and Morocco named Jawad, Nika and Vik, sister and brother from Kyiv, and Nika's boyfriend Vanya, art students who came here when the war broke out and who return often because they like the farm so much.

'On the 24th we filled our rooms with people, we accepted everyone, we put down mattresses everywhere,' explains Jolana. 'The 24th' is used as a date – without 'February' attached – that is deeply seared into the Ukrainian psyche as the date of the Russian invasion.

The refugees came from all over Ukraine. Some stayed a single night, on their long journeys to western Europe. Others stayed longer as they tried to plan what to do next. Jolana and Oreste organised transport, driving east to fetch people who needed to be brought to safety. In those early days and weeks of the war there was a sense that, even if Russian forces managed to overrun much of Ukraine, they might stop at the Carpathians.

Oreste and Jolana have two children of their own, young adults, working in France and Switzerland. 'If your son was at home now, would you encourage him to fight or escape?' I ask Jolana.

'I would say, you should live for your country, not die for it.' A country needs young hearts and minds, she continues.

Everyone should be able to contribute in other ways, without weapons. A gun even if it defends, still kills. And that's terrible. Not everyone is a soldier. On a state level, the Ukrainian state does not understand that. They need to buy weapons, and they need people to defend the land. But as a mother, I say: defend your land in a different way, as a computer programmer, as an interpreter. I am not a Ukrainian nationalist who says that everything Russian is bad, that every Russian should die. They're just people, too, and state propaganda in Russia is saying to Russians of eighteen or twenty: 'You must go now and fight for your country like a hero.' But if they die, their mothers' hearts hurt just as much as ours do. Here we are in the twenty-first century, and instead of putting all this money into science, into healing, into innovation and ecology, or into something new which makes human lives better, we are putting it into killing one another. How can we accept that? It's not about patriotism. I am a patriot of humanity, not only of Ukraine.

Western intelligence estimates at that point in the war suggested 120,000 Ukrainian soldiers had been killed, and perhaps 200,000 Russians, with at least three times as many wounded. The young people – especially the young men – of both countries are bleeding into the slow-thawing soil.

We tuck into a feast of pickled white mushrooms, beetroot soup, chicken stew, various versions of *ajvar*, pasta and pickled sweet cucumbers. The little white dog, now four years old, remembers me from my last visit.

Apart from the new faces, there are many subtle changes on the farm, due to the war. They have a new generator from Switzerland, in order to work the pumps that bring up water from their well, for themselves and the animals. Today only four hours of electricity is guaranteed.

My days pass in conversation with Oreste, Jolana and Jawad. Local people come to help Jolana with the animals. Oreste has Zoom meetings to organise his communal projects. The students draw, wash their clothes, or visit the refugees with Jawad and me.

Oreste has friends who volunteered and went to fight; but he knows local people who think it's all madness and who avoid the draft at all costs. His latest project is to encourage refugees who fled from the east to stay here permanently in Transcarpathia. This could be part of the answer to the rural exodus that preceded the war, he believes. His dream is to encourage them to work the land, to grow organic vegetables; but there's little market for organic food in Ukraine, and people expect to pay low prices for food. Suddenly, there is a new, unexpected workforce in the region. Oreste sees that as an opportunity, not a threat.

But there is a lot of resistance, locally. 'People don't speak out openly against refugees, when the whole country is united in opposing the invasion; but they mutter. They don't like them speaking Russian. They don't like their differences. And they – the refugees – are also uncertain who is going to win this war, and what the future will bring.' Unlike the people of Transcarpathia, the war has already burnt them. 'In our high mountains, there is less nationalism. People feel that any power, any capital city, is dangerous.'

We're sitting in their kitchen, while Jolana makes another meal. The biggest threat to Transcarpathia, as he sees it, is the same as it was two years earlier: the over-exploitation of the forests and the plans of the oligarchs and their political masters to build vast ski resorts and destroy the fragile ecology.

The media make much of the so-called crackdown on corruption in Ukraine, launched by Zelensky to prevent aid from the West going astray. 'Just window-dressing,' says Oreste, to impress the European Union and edge the day of Ukrainian membership closer. He shows me a press photograph, taken during the police raid on Ihor Kolomoisky's home. With his finger, he traces down the trouser leg of the police officer in charge. And stops at the ankle. The policeman is in socks! He and his fellow officers took off their big policemen's boots, in order not to soil the fat cat's carpet! How serious a raid was that? The crackdown on Kolomoisky and his businesses, he believes, will only be serious if they also go after his business partner Gennady Bogolyubov in London.

He produces a recent map that the anti-ski resort campaigners have managed to get hold of. The areas earmarked for development are blotches of red, scattered over the top of the massif where we walked two years before. The original project included 230km of piste; the new project has 700km.

The US and EU are already discussing how to rebuild Ukraine after the war. There are plans to build a circular road through the Carpathians, to open them up for tourism. And tourism will involve more ski resorts, building up to a future Ukrainian bid for the Winter Olympics. A place where the playboys and playgirls of the new Ukraine will glide effortlessly at high speed, while their brothers, or uncles, sit bitterly at home in wheelchairs.

Oreste and his fellow campaigners are fighting it in the courts. He lists transactions undertaken in the affected area by local mayors, who have bought up plots of land cheaply where the new resort buildings are supposed to be constructed. Those purchases are now being challenged. There are many contradictions here, too. He would like to lose his current case against the developers in the Supreme Court in Kyiv, he says, because he is impatient to take it to the European Court of Justice in Luxembourg, where he is more sure of victory.

I mention the Carpathian Biosphere Reserve. He waves his hand dismissively: 'Also corrupt.' Oreste is weary of patient explanation:

At the beginning, I wanted to explain what I am doing here with the farm, the restaurant, slow food, local produce. Now I don't want to explain any more. I'm fed up. They just want a big Mercedes and a big house.

There is also real poverty here. But people who own their own homes, who have a piece of land, have to think for themselves how to live better. And how not to exploit nature in the process. Instead, they're just destroying the forest. They want to destroy everything.

People ask me, what will we live off? If you're an activist, you have to offer some alternative. No, it is not my job to answer that question. Let them think it through for themselves.

After lunch, Jawad drives me down to the village to meet the refugees. The village council building has been converted into a reception centre. It's clean and warm, full of children's voices. Nastia, aged twenty-five, with two children – a baby girl and four-year-old boy – invites us into their room.

Her baby has Nastia's green-grey eyes. Her husband, Oleg, took refuge here with them, but he is in the army now: he was walking down the street in Khust when the police checked his identity. He was given three days to say goodbye to his wife and children; then he was gone, to the front. The big window looks north towards the mountains, and the floor of their room is strewn with toys.

Nastia was born in the Caucasus to a Russian father and Ukrainian mother. They moved to a small town near Zaporizhzhia for work, and fled when the missile strikes and shooting got too much. 'My son asks if we are bad people, because we had to run away. I tell him we are not. And that his dad has gone to kill the bad people, then he will come back.'

Valentin and Olga and their twenty-six-year-old son Yaroslav come from the Donbas, from a town occupied by Russian forces soon after 'the second war' began in February 2022.

'We had had war in our region for eight years already, but this was much worse,' says Olga. Valentin shows me a video on his phone – the block of flats where they used to live, riddled with shell holes, all the windows blown out, black burn marks scarring several floors. The video was shot from the back of a truck, on which cheerful men in camouflage uniforms can be seen. Russian soldiers.

'I felt I was Russian until the second war,' Olga says. 'Then I saw what the Russian side was doing.' So they fled west – though they have their reservations about the mentality of the people who have given them shelter. 'In our part of the country, people may be poor, but they share everything they have more easily. Here, people are much more calculating.' They are very grateful for the help they have received, but she can hardly wait to go back. Their son Yaroslav was cruelly treated back in the Donbas, and is mentally scarred by the war. He hardly speaks, and then only to those he trusts deeply. She doesn't think they will ever be able to go back to their home town. But somewhere close, maybe Dnipro. If need be, the easternmost regions could be traded with Russia, if it means a lasting peace, she says.

◎ ◎ ◎

I meet Márta for coffee in Mukachevo, during her lunch break from an international Catholic conference. Born in Budapest, her husband is from an old Transcarpathian Hungarian family. She moved here in 1997 and works for the Catholic charity Caritas, channelling humanitarian aid from Hungary to Ukraine. She's upset that the Ukrainian media ignore the Hungarian contributions, and only report the shipments from other countries. Hungarian Prime Minister Viktor Orbán is the 'second most hated world leader after Vladimir Putin', according to one public opinion survey, which doesn't make it easy to be a Hungarian in Ukraine. All four of their daughters are out of the country. All regard themselves as Hungarians.

Márta wants to be sure I understand the historical context. At Ukrainian independence in 1991, 80 per cent of locals voted for Transcarpathian autonomy within a larger Ukraine, and Hungarians wanted their own cultural autonomy, too. But it was not to be. The law-makers in the capital decided for a strong nation-state, rather than a federal state on the German model. Fear of future Russian attempts to break up the new state meant there was little appetite in Kyiv for decentralisation.

The various peoples of Transcarpathia lived at peace with one another until 2012, Márta remembers. She dates the 'national awakening' of the Ukrainians to 2014, when Russia seized Crimea and Russian-affiliated militias occupied parts of Donetsk and Luhansk provinces. She compares the Ukrainian awakening to that of the Hungarians in the 1840s. While the Hungarians defined themselves against the Austrians, the Ukrainians pushed back against both the Russians and the Hungarians. Language laws were passed making it harder to teach in Hungarian.

The conflicts over the language law blew up into a major issue between the two states. The Hungarian government blocked Ukrainian progress towards NATO membership. Ukrainians reminded the Hungarians of their own attempts to 'Magyarise' Ruthenian names in the years leading up to the First World War.

In the 2001 census, 151,000 gave their nationality as Hungarian. Under a law passed by the Hungarian government in 2011, people with Hungarian ancestors in neighbouring countries – including Ukraine – can apply for Hungarian citizenship, even though the laws of Slovakia and Ukraine forbid dual nationality. The law was warmly welcomed as an embrace from a long neglectful motherland. Many Hungarians from Ukraine went to Hungary to get their citizenship papers, which led to a red EU passport. As well as strengthening communities emotionally and economically, Hungarian citizenship ironically also acted as a drain on the Hungarian communities beyond the borders. For many, citizenship was just a stepping stone to a new life in western Europe.

After 'the second war' began in February 2022, Márta estimates there were fewer than 100,000 Hungarians left. She doubts that those who went away will ever return. 'They had more rights here as Hungarians in Soviet times.' Her brother-in-law took his university entrance exam to study in Uzhhorod in Hungarian in the 1980s. Nowadays, you have to take the entrance exam in Ukrainian, even if you aim to study at the Hungarian-language university in Berehove.

'In the Hungarian villages,' she explains, 'everyone listens to Hungarian radio, watches Hungarian TV, and conducts their everyday lives solely in Hungarian.' She sees the linguistic policies of successive Ukrainian governments as a deliberate attempt to push Hungarians out.

The strong Ruthenian identity adds another layer of complexity. 'Our next-door neighbour is Ruthenian, his son settled in Moscow. His other son was taken to fight in the Ukrainian army. They are on opposite sides now. Many families are split like that.'

Exactly who is a Hungarian is hard to quantify in a country of so many mixed marriages, and 130 different nationalities. She knows some Hungarians who volunteered for the army and went enthusiastically to war, believing they were fighting for Good versus the Evil of Putin's invasion. Others refused because they felt alienated by the language laws or because they never identified with Ukraine. Some who didn't flee at the start, thinking it would be over quickly, now live in fear of being forcibly drafted. They have created groups on Facebook to warn of military police controls. I ask to meet them, but they all say no. They are too afraid.

'Ukraine is a victim of both sides – Russia and America,' she insists. 'America sees the war as an opportunity to weaken Russia. We are a playground for the big powers, and I fear it will all end in great disappointment for Ukrainians.'

'In the end, those who cannot identify as Ukrainians will leave or have already left,' she says, 'and Hungarians as a distinct community in Transcarpathia will be lost, I'm afraid.' She takes refuge in her Christian faith.

I sympathise with her fears for her community, and mourn with her the disappearance of yet another colourful pattern in the multicultural mosaic of Austro-Hungary. But I can't help feeling that, with wiser leadership, all this could have been saved. Nationalism in one community generates nationalism in another. When the sparks fly, the smaller player must fall or flee.

I take a taxi to Rakoshyno (Hungarian – Beregrákos) to meet Jolana's brother Béla. I find him in the staff-room of the primary school where he teaches.

The back entrance to the school is protected by a wall of tidy white sandbags. Inside the entrance hall, the Hungarian red, white and green are next to the Ukrainian blue and yellow. The 'Rákos' bit of the name comes from the crayfish or river prawns that used to be abundant in the marshes and river here, the dried-up oxbow of the Latorica, which locals used to gather in baskets to sell in the market. There are still some left, but not enough to make much of a living from.

Béla teaches physics at the primary school, and is married to a Ukrainian. They speak Ukrainian at home. His wife's mother was a Hungarian Slovak, while her father is Russian–Jewish.

The teachers' room looks westwards. The sun sets swiftly, casting the shadows of the plants along the top of a shelf in beautiful patterns on the wall.

The local Hungarian identity is retreating, he says. There are villages with Hungarian-sounding names, but the population is solidly Ukrainian. Like Márta with her church, Béla has his own community, the Scouts. Through them, he feels at home in the Carpathians.

'We always go up into the hills to camp, usually to the last village in the valley. When we reach the last lamp-post, we go just a little bit further.'

It's getting dark outside. The horse's back – as locals call the nearest hills – has ridden out of sight among the mountains. Béla walks me to the bus stop on the main road. It's a starlit night, and the constellation Orion is visible above the southern sky, off hunting in Hungary.

After a while, I flag down a small bus, and I'm on my way.

CHAPTER 11

THE BEARS OF SYNEVYR

Time in the mountains is quite different. It is not to be compared
with foreign time.

Stanisław Vincenz, *On the High Uplands*

In a supermarket opposite the train station in Mukachevo, I meet
Serhii, our guide on the hike to Volovets two summers before, and
Julius, a friend from Budapest. We reported together on the wars in the
old Yugoslavia in the 1990s. He went on to cover conflicts all over the
world, then escaped and retrained as a bear guide in British Columbia.
He has a plan to take injured Ukrainian veterans back to Canada with
him, to retrain as wilderness guides.

Serhii's wife, Vika, serves a breakfast feast of aubergines in sauce,
tomatoes and beans, eggs galore, mashed potatoes, meats and cheeses,
pickled cucumbers and cauliflower. Their daughter Nastia and sons
Taras and Zakhar, just one year old, circle around. Zakhar has a big
voice and smile, and looks like a young version of Serhii. Serhii is
excused military service because he has three children, and can even
travel abroad.

Viktor Stenich, another of the hikers, comes to Serhii's house to
meet us. His foreign travel business evaporated when men could no

longer leave Ukraine, so he began organising food boxes. He has a thousand customers a month, 10 per cent in Transcarpathia, 50 per cent in Kyiv and 15 per cent on the various war fronts. There are two sizes of box: the smaller is $10 and the larger $23.

Inside there is cheese, meat and sometimes special treats, like a bottle of good wine. 'People long for a taste of their old lives, before the war,' Viktor says. 'Our guys on the front need to know that someone remembers them.'

Hundreds of thousands of men and women have been traumatised by the war. Half a million are serving in the Ukrainian armed forces. Tourism, Viktor acknowledges, will never be the same again.

Serhii produces the Maramures Ecological Network for Brown Bear map, showing the main habitats and migration corridors for bears across the Ukrainian–Romanian border.

'If you train veterans now,' Viktor says, 'they will forget what they learnt after a few months.' Why not train existing mountain guides like Serhii in nature therapy skills, to add to what they already know about their own mountains?

Julius wants those who come to feel they are teaching the Canadians something, too, for their own self-respect. They should not feel they are recipients of charity. British veterans from Afghanistan teach surveillance techniques, which could be adapted to watching bears.

We set out from Rakhiv early, to visit the bears in the Rehabilitation Centre in the Synevyr National Park. There are few wild bears in Transcarpathia – only around 400, compared to more than 6,000 in Transylvania. No respecters of national borders, they migrate to and from Romania, in search of food or a mate. Many bears in Ukraine have been trapped for state and private zoos. Others made their way into eastern Ukraine from Russia – the archetypal 'Russian bear', perhaps. The war has been even more perplexing for bears than for humans.

It's a journey of several hours, back to Khust then up into the mountains. We make a break in Iza, famous for its baskets, made from the willows that line the shore of the river Rika. From the sheer number for sale, up and down the main street, Iza may be satisfying the basket demand for half of Europe! There are twenty kinds of willow, they say.

In the largest store, we find willow rocking chairs, lampshades, rocking horses and willow-woven paraphernalia of all shapes and sizes, from place mats to giant log baskets. Most are natural brown, others painted white. Ivan Senevich, the son of the owner, says business is bad, but at least the baskets keep.

He has a huge selection of axes and axe heads decorating the walls. People bring the axe heads when they find them in their fields. Some they give, some he buys. Each iron axe head is a slightly different shape, for different ways of splitting or forming wood. Some date to the Iron Age. The oldest looks like a blob of crude iron with a hole for a wooden pole. There are even stone cudgels from the Stone Age. There are enough axes on the walls here to equip a small army.

The road runs up the river valley beside the Rika, towards Mizhhir'ya – Ökörmező in Hungarian, meaning 'ox-meadow'. The abandoned concrete bunkers of the Árpád defence line runs through here. Everything is buried waist-deep in snow.

We drive through Horinchovo, a village famous for its bandits and storytellers. Its name means 'from burning', because the authorities came to set fire to the thickets where highwaymen were believed to be hiding. The twentieth-century folklorist Petro Lintur lived here and collected fairy stories and legends throughout Ukraine, especially from the Carpathians. In Horinchovo he found storytellers with some of the best memories and tales. As a convinced Marxist in Soviet Ukraine, Lintur placed the folktales in the East Slavic tradition, in close relationship to Russian and Belorussian tales. But he was a good enough storyteller to acknowledge the western influences, especially on the western slopes of the Carpathians. He divided his collection into animal tales, tales of magic, legendary tales and tales of everyday life. He found much to fit his ideological leanings. The peasants always had a sharp eye for injustice and the abuse perpetrated by cruel landlords and greedy clergymen. Each summer in Horinchovo there is still an annual festival of storytellers.

Mizhhir'ya is a bustling hill town with a war memorial of a newly gilded statue of a soldier, and another of a shepherd blowing a *trembita*. We turn right onto a road which leads over the Synevyr saddle, then down to the administrative buildings of the national park.

Mykola Derbak, director of the Bear Rehabilitation Centre, wears a green hunter's fleece and radiates authority. The Tereblya river bubbles noisily outside, past another Árpád Line bunker.

The staff are preparing for the wild animal census, to track bears, deer, wild boar, foxes and wolves. The snow began falling a couple of days earlier, so the tracks will be fresh. In the old days, it started in November at the latest. One of the foresters saw a mother bear and a cub the previous day. The mother bears weigh around 200 kilos, the males up to 300.

The park authorities put out salt for the deer. There are no incidents on record of wild bears harming people here, though there are frequent encounters, especially in bilberry season. They have thirty bears in the reserve, mostly rescued from private zoos. They are free to roam round a wooded reserve of 12 hectares, with a cliff, thick forest and two streams. Mykola's organisation has identified another 190 captive bears around Ukraine which could benefit from this freer environment, so he wants to expand the space, if he can raise the money. Two of the bears are blind.

It's feeding time and more than a dozen bears are clustered just behind the fence, chewing on frozen fish. As omnivores, bears eat what they can get. Meat could make them aggressive, so it is not on the diet. The oldest bear is thirty-five. In the wild they rarely make it above twenty. Two cubs have been born here, but one was eaten by a large male, jealous of the attention the mother was giving the little one when he wanted to mate with her. The other cub fell from a tree and died.

Poaching is still a big problem in the Carpathian Biosphere Reserve. There's a shortage of rangers, as many are away, fighting on the front lines. Another problem is the lack of telephone signal over much of the national park. And the rangers' night-vision binoculars have been transferred to the war zones.

Julius is surprised to find out that, even here in semi-captivity, the bears behave much as they do in the wild:

> If you watch them carefully, you see they're figuring out their dominance without actually fighting – those little movements, occasionally a little growl. But usually, it's just head movements. See the one on the left? The mouth movement. That's low-level stress.

Another one approaches.

> So he's coming in, but I've got a feeling the one on the right is more dominant. So let's see if he even allows him to get close. See that? He moved away. And that guy moved away slightly. So this is a continuous kind of game, a dance.

He's also surprised how different each of the bears looks – not just the different sizes, but the coats, some of which are much lighter than others. In British Columbia, the bears are more similar. After the age of eight, it's hard to tell from their size alone how old they are. Other distinguishing marks to look for are scars from old fights.

> People think bears have bad eyesight. It's not true. They have quite good eyesight, but it's slightly different from ours. They're not very good at detecting complex shapes against a complex background. So you could be sitting on that bank over there and if you were in trees or bushes they probably wouldn't see you.
>
> Typically, you wouldn't see two bears within 50 yards of each other, unless there's a concentrated food source. And when they come closer, they sort of navigate their way through it. Bears that spend a lot of time close to other bears are safer, because they're

better socialised, they have better understanding of bear to bear dynamics. And when they meet a human, they do exactly the same thing. So if you understand what the bears are doing with each other, you know what the bear will do with you as well.

◎ ◎ ◎

The little, crowded bus makes slow progress on the potted road, parallel with the Tysa river, burrowing deeper and deeper into winter. At Tiachiv the bus stops for a break, and Serhii gives us a brief tour of the town. By the main road there are photographs of the old bridge over the Tysa, across to the rounded hill on the Romanian shore. The same hill appears in the work of the Hungarian impressionist painter Simon Hollósy, with steep-roofed peasant houses in the foreground. There is a bust of Hollósy beside the Calvinist church. He was born just across the river in Sighetu Marmaţiei, and spent the last part of his life in Tiachiv. The bridge has not been rebuilt.

There are memorials to three wars and one popular uprising, side by side. One to the dead of the First World War, on which nearly all the names on the black marble slab are Hungarian, in Latin letters: Sámuel, József, Gábor, János, Bálint.

Then the Second World War, this time in Cyrillic letters, but again mostly Hungarian names: Imre, Károly, András, Sándor. Above the monument, a stone imperial eagle looks upriver, and a black Austrian one looks downriver.

By the time of the Afghan war in 1979, the names are all Ukrainian or Russian: Yuri, Vasyl, Oleksandr, Iosip, on dark-red marble. The white outline of Afghanistan, with the main cities marked, is drawn above the names. In front of the monument, is a charcoal-grey triple cross.

In Cyrillic, the name Afghanistan seems strange: Афганістан. If it was Russian, it would be written with one letter different: Афганистан. To its right, engraved in the stone, are two soldiers, one standing and the other wearing what looks like the hat Serhii wears – to tie in the contribution of Transcarpathians, perhaps.

Another monument, cast in bronze beneath a white stone archway, shows a cluster of people, shouting or singing, while the foremost, moustachioed man has his hand on his heart. Beneath their feet, the bricks of a wall are crumbling. This is a monument to the Maidan revolution from November 2013 to February 2014, which toppled President Yanukovych and began Ukraine's turn towards the European Union and NATO. The protests began when the government suspended the signing of an association agreement with the European Union, to seek closer economic ties with Russia. That fateful turn towards the West, encouraged by the United States, led directly to the current war: President Putin of Russia could not tolerate a westward-leaning Ukraine.

On one side of the monument is the blue-and-yellow flag of Ukraine, symbolising wheat beneath a summer sky; on the other, the red-and-black flag of war, symbolising blood and destruction.

There is also a gold bust of a Soviet-era airman, and another of the nineteenth-century leader of the Hungarian war of independence, Lajos Kossuth. All the martyrs and heroes are clustered here: you can take your pick, to whom you bow your head or place wreaths of fresh flowers.

On the road to Rakhiv, our bus is stopped regularly at military checkpoints.

'All passengers locals?' the officer asks the bus driver, kindly offering the correct answer to the question. As British passport holders, Julius and I are of little interest to the Ukrainian military. They are only curious about Ukrainians from elsewhere, finding their way close to this westernmost border. This is a favoured route for young men trying to escape the draft. Coils of razor wire stretch along the roadside, just as in 2021. But what was designed to stop smugglers taking cheap cigarettes to Romania is there now to stop male Ukrainians fleeing for their lives. Swollen by snow and ice, the Tysa presents a frightening prospect to swim across; but it is a long river, with nearly 200km of border with Hungary and Romania. There are always old tyres and small boats.

The nearer we get to Rakhiv, the deeper the snow. We book into a guesthouse run by Péter Popovics, round the back of the Hotel Europa. The rooms are all pine and paintings, with a map of 'Transcarpathian Rus' in Czechoslovak times in the entrance hall, and broad wooden skis from the same era. Upstairs, there's a jug of strong red wine on the table to welcome us.

Péter is Hungarian, a tall, serious man, who drives frequent sorties to the east of Ukraine to take supplies. He was born in 1970 into a communist family, and remembers the bread queues of the 1980s as the first sign of the weakening of the Soviet Union. 'It all began after the Summer Olympics,' he remembers.

'Nothing in the shops, everything we needed at home,' a common survival story of eastern European families under communism. Ukrainian independence in 1991 brought hopes of prosperity. 'We thought Transcarpathia would soon be part of the West . . .'

Even in the snow – or because of it – Rakhiv feels like Romanian towns did thirty years earlier: the same faces and smells, the flaking grandeur, sturdy Lada cars skidding on the ice. Péter reckons the most prosperous period was under President Yushchenko (2005–10), 'though God knows where the money came from . . .'

We drive together up the valley of the White Tysa. I wade through knee-deep snow to photograph the point where the White and Black Tysa merge. The 'white' name comes from the marble-like stones visible in its waters on a summer's day. Today, the white shores encroach on the river, and the trees are heavy with snow.

We pick up Ahava at the mayor's office in Bohdan. She is a sprightly seventy-eight-year-old, wearing a woollen jumper in two marvellous shades of blue. Her name is of Hebrew origin and means 'love'. Péter has arranged for her to take us to the home of her mother, Anna, aged ninety-five. The snow-beaten track leads up past the village church, through a garden gate. Anna stands in the doorway to welcome us, with that humility of the very old who insist they don't understand why anyone would be interested in them or their memories anymore.

Her cottage has two rooms, and she spends most of her time in the living room, which is also her kitchen. There are icons on the walls, a church calendar, a wood-burning stove and a crowded sink with a single tap.

'I was last in Budapest in 1941,' Anna tells us. 'I went there with a load of wood, by train, with my father.' She would have been thirteen then. If we had more time, I would ask her where they stayed, what they did and what the city looked like in the second year of the Second World War. But we have so many questions.

'What about Moscow?' Julius wants to know.

'Once in 1958. Then in 1966. I had two days there, shopping, on the way back from Siberia.'

Anna's first husband fought in the Hungarian army in the Second World War. He was taken prisoner by the Russians and caught tuberculosis in captivity. He came back a broken man, and died in 1949. Ahava was born in 1948, and has only the vaguest memory of her father.

In 1952, Anna married again. The new husband, Ahava remembers, was a good man and treated her like his own daughter. He bought her clothes, shoes, anything she wanted. He died in 2002, in Altai, in southern Siberia, where he had gone to cut wood as a forester, aged seventy-five, to save money to improve their house. It sounds strange that someone from a village with so many trees should travel thousands of kilometres to earn money chopping down trees. But I remember young Ukrainian soldiers I met in the military hospital in Kyiv who had gone to Russia to work on building sites, because the pay was better in Russia. A few weeks later, they found themselves on the front lines in Luhansk and Donbas, fighting young men they might have been working with, shoulder to shoulder in Moscow.

Anna's first memory as a child in the 1930s is of gathering twigs in the forest, while her father cut trees. Her grandfather was Hungarian and her grandmother Ruthene. Her father spoke Hungarian and Czech. In those days, many spoke German too – and 'the Jewish language', by which she means Yiddish. The fire crackles constantly in the stove, and the water inside the kettle comes slowly to the boil.

Our questions are disjointed, spoken in Hungarian or Russian, translated into Ukrainian by Péter, then translated back into Hungarian. So we gather fragments of information.

She remembers her mother telling her how Romanian soldiers occupied the village in 1919, stole everything and treated people very badly. The years of the terrible famine from 1932–3, known in Ukraine as *Holodomor*, 'death by hunger', were not as bad in this valley as 1947, she remembers, when they had no bread at all. In 1935, there was a big strike and the Czech police came and beat them, and they fled into the forest.

The Holodomor was a man-made disaster, imposed on Ukraine by Stalin in response to an attempt by Ukraine to achieve more autonomy within the Soviet Union. The main victims were rural Ukrainians in the great grain belt which stretches across the centre and south of the country. At a time of a new Russian invasion, the memory of the suffering of the 1930s seems especially poignant.

In 1938, many people emigrated to America, others to Russia. When the Soviets took over Transcarpathia and the hunger of the late 1940s was over, life was not too bad, Anna remembers. She was young then. 'As a woman, you could even go to university, if you wanted to.' And there was work in the cooperative. She met her second husband, got a job in the village library, and worked there all her life, retiring only nine years ago, aged eighty-six.

Ahava shows us an embroidered white smock and sheepskin waistcoat, traditional dress for women. The embroidery is beautiful, with a green thread interwoven with all the reds and yellows and blues that go to identify a Bohdan waistcoat. It is cut in a special way, with an opening at the front and back, so you can wear it on horseback.

'The Jews used to come and ask if we had any sheepskins to sell,' she says. The Jewish girls and the Ruthenian girls would sew the patterns. She had Jewish friends at school in the 1930s, but they were all taken away in April 1944, during the Easter holiday, to the concentration camps, and never came back.

Ahava constantly fills our glasses from a bottle of Chivas whisky, which may have been sitting in her cupboard for many years. She and Anna also drink a little. And when the kettle boils, she pours us strong black coffee with lots of sugar, and insists we try the cakes she brought specially from Rakhiv. 'Eat something,' Ahava says. 'Everything you can see you must eat.' The iron rule of Hutsul hospitality.

Anna sings little ditties, which sound impossible to translate, so nobody tries. Ahava's two children and two grandchildren live in Slovakia. Her son Ivan studied physics in Russia, and qualified as a teacher, but works at a ski resort now. 'But as there's no snow any more, he does other jobs.' Yuri is four, and the youngest grandchild, Nina, just four months. Anna hasn't met her great-granddaughter yet, but has seen her picture on the telephone.

Ahava wants to know if we have heard of 'Sarban', the most famous son of this valley, who won an international tree-cutting championship in Canada in the 1970s. He could cut down a tree with an axe and make it fall in exactly the right place, faster than anyone else in the world.

Our glasses are refilled and we clink again. The ladies are getting tired, so it's time to go. I squeeze in a final question. Who are you both?

After a little thought, Anna says she's a Hutsul from the Bohdan valley, while Ahava says she's a Hutsul from the Breboya region. The older you are, I note, the narrower the place you identify as home. We bid them a fond goodbye, and stumble back into the snow. The sun is shining, and all the dogs of the valley seem to be barking in unison.

◎ ◎ ◎

Thirteen men trudge up the track from Bohdan, just across the valley from Anna and Ahava's house. Most are in their twenties and thirties, but Liviu is in his mid-fifties, an ethnic Romanian from Ukraine. He's with his son, Petru, the only one of the group with real experience of the mountains – he used to be a snowboard instructor. It's April 2022, a few weeks after the full-scale Russian invasion of Ukraine. I have changed all the names, to protect their identities.

The men barely speak to each other. The way will be long and the track is steep, and each is thinking he must conserve every gram of energy, every breath. The only sound is the thump of their feet on the stones, a few frightened birds and the trickle of a mountain stream, unseen among the trees.

After a while, snowflakes start to float down around their heads like moths, or spies sent from the top of the mountain to identify intruders. The older man stumbles and curses. The snowflakes thicken and settle on their coats, their eyebrows. The scene would be beautiful, if it was not so desperate.

These are men fleeing their country in wartime, because they do not want to fight. Their motives are mixed. Some don't want to risk their lives for a Ukraine they don't believe in. Some have been convinced by their loved ones; others needed little convincing.

The snow softens the tramp of their feet as they march, reducing the chance of detection, but increasing the danger. Everything is quieter now, and their footprints will soon disappear without trace. The border guards smoke lazily in their barracks.

The hearts of the men on the mountain sink. They had hoped for spring, but are not surprised. These Marmaroshchyna (Maramureş) mountains have a reputation – and not for gentleness or consistency. Through the trees ahead of them, they catch glimpses of the line of peaks that they will soon have to cross.

The sight causes Liviu the most distress. Did he talk his son into this, or did his son persuade him? He can't remember anymore, and it is too late to ask. Perhaps it was his wife who suggested they go together. From the warmth of their flat, he imagined that he was going along to look after the boy. As he watches his son now, striding ahead, deep in conversation with one of the smugglers, it crosses his mind that the boy is here to look after him; that their roles have been reversed.

After a few hours, they call a halt. They look each other over, a wretched sight. Their coats are soaked through, few have boots good enough for a long hike through treacherous mountains in late winter. Some of their gear was bought for a pittance at the second-hand clothes store in Rakhiv, discarded items from rich tourists. If it was waterproof once, it is not anymore. Each of the men carries a knapsack with food: rough white bread and cuts of meat and white cheese, lovingly prepared by the womenfolk.

Dima wears a silver necklace with big links and his name written on the back, a lucky charm. He's thicker set than the others. In normal life, he might be an estate agent or a lawyer.

Only the smugglers have proper footwear, Dima notices, and they will soon turn back. He was born not far from here, in a town by the Tysa; but after half a lifetime away in Russia, building houses, he doesn't know these parts well. Some of the boys are locals, others come from much deeper inside Ukraine. Each has paid between €2,000 and €4,000 for this trip. He got it cheap, because he knows one of the smugglers from his schooldays.

What had they actually paid for? Two or three months' wages to be shown this track and a few dotted lines on a phone? To be driven here at the right moment, when no patrols on the lookout for draft dodgers like him might cruise round the corner in military jeeps? Some of the money probably went to the border police, to turn a blind eye, but most stayed in the pockets of the shark-faced smugglers. He recognised them from his school playground. Hustlers, dealing in cigarettes, even then. Now they hold his fate in the palms of their hands. The thought makes him forget to breathe.

The price contains no guarantee of success. They shelter from the snow in the lee of a foresters' lodge. They discuss resting for the night, but they are only three hours out of the village, and the smugglers insist they go on. The hut is too exposed, too obvious a place for the Ukrainian border guards to look, if they choose to. Reluctantly, they leave the shelter to plunge back into the wind and snow, the encroaching April night.

Could they not have waited another month? Dima wonders. True, he had received his call-up papers only two days earlier. It was a Sunday, and the next morning, at eight o'clock sharp, he was supposed to report to the military office in Uzhhorod. He imagined the men standing, heads slightly bowed in the courtyard, and the silence when his name is called out, then called out again, and the men looking up, and the glances exchanged, as the officer bites his lip then continues the list.

He thinks of his wife, Katja – like him, from the Romanian minority in Ukraine. It was actually her decision that he leave. Until the very last minute, they didn't tell the children – a girl of nine and a boy of twelve – that their father was going to cross the mountains into Romania, so that they would still have a father. That once he made it, they, too, would leave their home, school, friends – everything they knew – to join him in exile in another land. But they all have to survive this war, he tells himself.

Another break. The wind getting fiercer, no let-up in the snow. The smugglers smoke the most. Dima doesn't smoke. Nor does Petru. The older man coughs painfully. Shouldn't he have stayed behind in the valley?

They know the border runs along a sharp ridge, followed by an almost vertical drop on the Romanian side. The peaks are lined up somewhere in the storm ahead. Pip Ivan, the highest, just shy of 2,000 metres. As a child, he thought it was a person: Ivan the Priest.

The worst part, the smugglers told them, days ago, over coffee and cigarettes, will be when they are on the ridge, already safe from Ukrainian army patrols, but at the mercy of the wildest section of the mountains. No smugglers there to guide them. The range on the far side of the border is almost uninhabited for the next 30 or 40km, with no tracks other than those made by bears and wolves. Dima is afraid of the bears, in particular. They seem a more tangible threat now than the steep, trackless way down the mountain.

As darkness falls, the smugglers turn back. No handshake, no team spirit. A gruff 'good luck', and they are gone. He envies them, making money from this. Now he and his people are on their own.

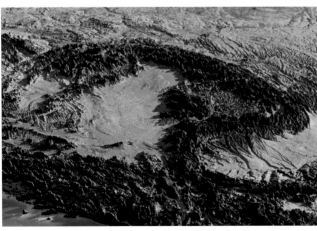

1 & 2. This artist's rendering of Europe's topography illustrates how the Carpathians anchor these lands at the geographic centre of Europe, forming the watershed between the Baltic and the Black Sea. They pass through seven countries: Austria, Hungary, Slovakia, the Czech Republic, Poland, Ukraine and Romania. I follow the great horseshoe of the mountains clockwise, 1,500km from the River Danube near Bratislava, to the Danube at Orşova, on the Romanian border with Serbia.

3. My journey begins at Devín, in present-day Slovakia, where a cliff topped by a ruined castle slopes dramatically down to the Danube. The Árpád monument is clearly visible. It was built to celebrate 1,000 years of Hungarian rule in 1896, and blown up in 1921, by which time the hill was part of the new state of Czechoslovakia.

4. A Hutsul family. The Hutsuls are one of Europe's lesser-known and most intriguing ethnic groups, living in the Carpathian highlands of Ukraine and Romania, so old they have their own creation myth of the beginning of the world. They are famous for their skill at embroidery, woodcarving and music.

5. Hunting party of Count Lajos Königsegg, Transylvania, 1900. The Carpathian forests are home to wild boar, deer, wolves, lynx and bears. How much land do wild animals actually need? How much access should shepherds and woodcutters, hunters and tourists have to the remaining pockets of Europe's last wilderness?

6. The lost island of Ada Kaleh. The island was a marvellous remnant of the Ottoman empire, famous for its fig jam, Turkish Delight and pistachio ice cream, lost to the damming of the Danube in 1968. The river divides the Carpathians from the Balkan mountains.

7. The High Tatras in Slovakia. Only 75km long, they include the tallest peak in the Carpathians, Gerlachovský štít, at 2,654m. This picture shows the Tichá and Kôprová valleys on the southern, Slovak side, looking towards Poland. Environmentalists won a long battle with the Slovak state to grant them maximum protection. There are stone pines at least 700 years old.

8. On the *polonina*, the high uplands of western Ukraine, emerald meadows brimming with purple irises and other wildflowers. The Chornohora ridge, crowned by Hoverla peak, rises up to the west. This photo was taken on my last day in the Hutsul lands, in June 2021. The cheeses in Plate 13 come from the milk of cattle which graze on these pastures.

9. Svydovets, an almost untouched massif in western Ukraine, is threatened by the planned construction of one of the biggest ski centres in Europe. The value of good drinking water and timber is high, but local communities here feel deprived. Many people are tempted by the oligarchs' promises of instant wealth.

10. Lake Ștevia in the Retezat mountains is a fierce grey-green, like whipped glass. After a difficult approach over huge boulders, under a threatening sky, I crouch at the water's edge, wash my face in the cold water and cup my hands to drink a little from one of the 'eyes of the sea'.

11. Anna, ninety-five, in her little village in the eastern Carpathians, beneath Ukraine's highest peak, Hoverla, describes herself as a Hutsul from the Bohdan valley. Her daughter Ahava, seventy-eight, says she's a Hutsul from the Breboja region. But the war with Russia is encroaching on this most remote corner of Ukraine.

12. Denis's forearms are stained black by the coal dust. His father drove the same locomotive, and his grandfather before him, through the Vaser river valley in the Maramureş mountains of north-west Romania. There are so many levers to pull and adjust, it feels like being inside the engine of a giant motorbike.

13. Vasyl speaks in the loud voice of someone who spends most of his life outside, bellowing over the wind. His cheeses are made from cows' milk, strained in muslin over the wooden pails in his outhouse. He went abroad once, by mistake, when he strayed through the forest on the Ukraine–Romania border.

14. In the middle of the feast stands a bottle of home-made *horilka* with a deliciously bitter taste, from stinging nettles. Vasyl flourishes a big bunch of them. After a lifetime carving flutes and mountain horns, they cannot harm his big, chiselled hands. In the eastern Carpathians, in western Ukraine.

15. *Daybreak in the Retezat Mountains* by Andrea Weichinger. The roar of the stream, the countless stars framed by the mountains, the absence of light pollution – those are the real luxuries.

At this point, the mountain-crossers have their first row. It is as if the smugglers had been the teachers on a school excursion, and now they are out of sight the kids start to fight among themselves. Some argue for staying in the lodge overnight. It would be madness to continue in the dark and the snow. Dima, Petru, his dad and another man retort that the smugglers may be caught by an army patrol as they return to the main road. Then soldiers will be on their trail. Better to get the worst part of the journey over tonight, and get up almost to the ridge. These four men split off from the others and go on ahead.

It is a wretched decision. After a couple of hours, they give up and go back to the hut. The others watch them arrive, sullenly.

First thing in the morning they try again, just the four of them. Suddenly Liviu cries out and falls out of sight. He has tumbled down a ravine. He waves a dismal hand from the bottom: his pride is hurt, but no bones, he thinks. Petru begins a long, treacherous descent down the side, to rescue his dad. It takes him an hour to get down, two to get back, half carrying his father on his back. With a penknife, he cuts footholds in the soil on the side of the ravine, a superhuman effort. The others watch in silence.

They carry on, painfully slowly. As darkness falls, they make a hole for themselves in the undergrowth and lie as close as they can to one another to keep warm. At one point, Dima notices Liviu, the old guy, get up and start walking away. 'Where are you going?' he calls out, over the roar of the wind. 'The kitchen . . .' Liviu shouts back. At that point, Dima realises Liviu is losing his mind. He wakes Liviu's son, and the two of them get hold of Liviu and bundle him back into their huddle. When they wake in the morning, Liviu is missing – this time for good. This is the Monday night, their second on the mountain. Their food is running low – they brought only a little, to minimise weight. Romania looked so close on the map. Dima feels wretched. The wind howls and the sleet stings his face. He sits in the snow while the others rest. He has just enough signal to ring Katja. 'I won't make it,' he tells her, in tears. 'Forgive me.'

'Don't be stupid,' she says. 'If you love me and the children, you have to.'

He finds himself standing on the ridge, next to a pole which marks the frontier. They are safe from the Ukrainian border guards, but not from the mountains. The next moment, he loses his footing and finds himself flying down the mountain, bouncing on snow drifts packed hard by the ice. At moments, he is flying through the air. The thought crosses his numb mind that if he hits a boulder or a tree at this speed, he will be killed. He tries to claw his fists, his knees into the mountain-side, to slow his descent. At last, he feels himself stop. Deep in a snowdrift. Is this death? But surely the dead don't feel pain?

Miraculously, he is badly bruised, but nothing seems broken. The worst thing is, he has lost both his boots and one of his socks. And his phone. But he is alive, in Romania, on the right side of the mountain.

Dima tears a piece of his trouser leg to improvise a sock for his left foot, and ties it in place with his phone-charging cable. Then he carries on, down the mountain. The snow is worse this side, and he falls often. He tries to follow a stream. At one point he looks up and sees what looks like a church spire. He vaguely remembers a church on the map he studied with the smugglers. A church means people, and rescue. No longer able to feel either of his legs, he half climbs, half crawls up to the place. But there is no church. He has been hallucinating.

He stumbles back down the mountainside to the stream, and falls badly again. His left heel is a mangled mess, bleeding profusely. A mortal terror of bears engulfs him. They can track him easily, from the trail of blood. But the pain is too much. He cannot go on. He hauls himself out onto a rock in mid-stream, curls up and waits. Here at least, the rescue services, if there are any, might see him, and he will see any wild animals coming for him.

At some point he hears his name shouted. Petru has found him! Petru's father is gone, and he had lost the other two men on the way down the mountain; but he had found Dima! Petru presses on, down the mountain, to look for help.

After her husband's phone call on Monday night, Katja does not waste time. The children have told her: 'Don't come home without daddy.' She drives to the bridge at Solotvyno, parks the car and runs across. Her Ukrainian passport has expired. The border guards shout at her, but let her go. Seeing a crazy Ukrainian woman running towards them, the Romanian border guards try to block her way. She quickly explains the situation to them. They take her into the border guard's office and call the chief, Adrian. In her presence, he rings the mountain rescue service. The snowstorm is raging, even here in the river valley. In such conditions, the Salvamont people tell them, it is too dangerous to put up a helicopter. She should try to tell them approximately Dima's last known position. She rings the smugglers. Reluctantly, they give her some map coordinates. Then some others, when it turns out those are false. The Romanians say none of them are true. All this time she is biting her nails in a tent, turning every few minutes to cajole the border guards for information. When night falls on Tuesday, she realises there is nothing more to be done, and walks back across the bridge into Ukraine. The young border guards let her back in, even without documents. Then she drives to her cousin's house. She cannot face her children.

The next morning, Adrian, the Romanian border chief, rings her. They've managed to put a helicopter up, he says, but have found nothing.

Katja takes a desperate decision. She goes to the Ukrainian border police post in Rahiv and confesses everything. Please help find my husband! She has no way of knowing that he is close to death that moment, on the Romanian side.

The officials listen sullenly. Then her cousin rings to say that the Ukrainian police have caught six men on the mountain, and that they have been taken to the barracks. She drives there frantically. The gates are locked and guarded. Other women arrive and start shouting at the guards to stop beating their husbands.

At midday on Wednesday, Romanian television – which can be watched in Ukraine – reports that three men have been found alive, and one dead, in the mountains. Katja doesn't know whether to laugh or cry. Adrian, the Romanian border guard, rings again. What distinguishing clothing was Dima wearing? She tells him about his heavy silver necklace.

On Wednesday afternoon, Dima sees two men in red jackets clambering up the stream towards him. The Salvamont. From a distance they wave, showing a thumbs-up sign. He gives them the thumbs-down. He is barely conscious.

Once they reach him, he gets a sip of water; he's been eating just snow for two days. Then a square of chocolate. He retches.

A helicopter hovers overhead and a harness descends. He is winched to safety. An hour later, from the back of an ambulance, a nurse lends him her phone. He rings Katja.

'Imagine baby, I made it!'

A year later, in the spick-and-span office of a charity in Baia Mare in northern Romania, Dima rolls down his socks to show me his left foot. Just a stub remains, like those of the lepers I once saw in the leper colony in the Danube delta. He lost all his toes to the frostbite.

'Do you have any photos from the rescue?' I ask.

That proves to be another story. Katja went back to meet the smugglers a few days after Dima was rescued, to demand their money back. In their anger, they grabbed the phone from her hand and smashed it.

'Will you ever go back to Ukraine?' I ask Dima.

'I would like to. Katja and I, and the children, we miss everything about our lives there.' The war is another year older, 100,000 more men have died, fighting for their country. What would he say to their relatives, if they tell him he is a traitor, I ask as gently as I can.

'I have nothing to say. For me, my family matters more than my country,' he replies.

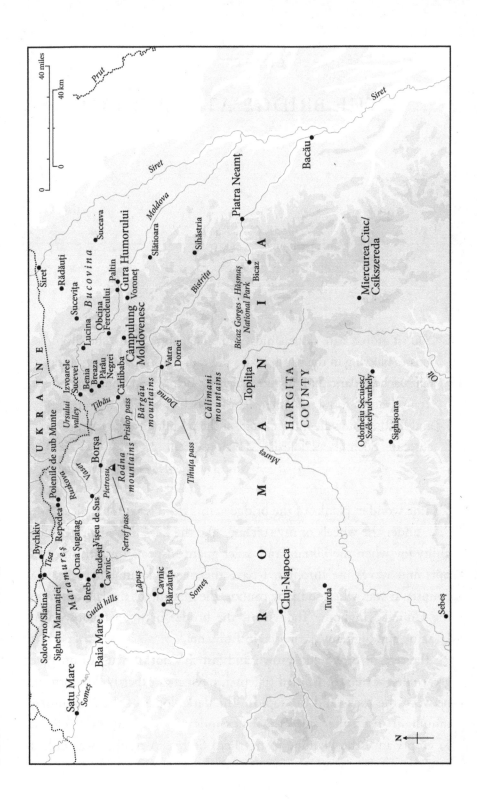

CHAPTER 12

THE BRIDGE AT SLATINA

You sit atop a bare ridge amidst unfamiliar mountains, the silvery
grasses blowing in the wind. Like a polecat, the Carpathian evening
approaches, dark and furtive. Silence. Below you, nothing but
Transylvanian forest, no other sight to catch your eye. Green ridges
and foothills flow like streams out of the mountains, the solitude is
absolute. Such places do exist. Which way will you go?

Miloslav Nevrlý, *Carpathian Games*

The wooden planks of the bridge across the Tisa rumble pleasantly
under the wheels of my suitcase, like the keys of a piano. I feel a
guilty joy when the Ukrainian border guard inspects my British passport and waves me through. I have the magic document to leave a
country at war, when so many cannot, the blessing of the late queen of
the United Kingdom. The Tisa is the border between Solotvyno in
Ukraine and Sighetu Marmației in Romania.

The river is broad and brown and fast. It's not difficult to imagine
great rafts of tree trunks tied together, navigated expertly downstream
between the rocks and the rapids. The dark slopes of the Maramureș
mountains rise either side. The Tisa winds and loops nearly 1,000km
to empty into the Danube in northern Serbia. A donkey was tasked

THE BRIDGE AT SLATINA

with ploughing a furrow for the river, runs a Hungarian legend. At first it went straight, then was increasingly tempted by the delicious purple thistles on either side, and meandered this way and that.

The war in Ukraine has turned this bridge into a meeting point for those fleeing the war and those returning to it. It looks old, but only opened in 2006, cobbled together from parts of another bridge further upstream that was blown up at the end of the Second World War.

The first impression crossing on foot from Ukraine to Romania is the sound of dogs barking. The people crossing with me are mostly locals, elderly folk. Some 150,000 Romanians live in Ukraine, according to the 2001 Ukrainian census – the same number as ethnic Hungarians; meanwhile 50,000 Ukrainians live in Romania. The Tisa feels less like a border river, and more like a torrent flowing through the western fringes of Ruthenia.

If not for the war, this would be a dreamy eastern European backwater, abandoned by the young and cherished by the old. Pedestrians are channelled down the left side, away from the cars. The bridge is not robust enough for trucks. In the queue, I chat with a Hungarian-speaking couple, residents of Solotvyno, clutching shopping bags. Young men who want to escape Ukraine, they tell me in hushed tones, usually succeed 'one way or another'. Many local pensioners deal in cigarettes, which are €2 a packet cheaper in Ukraine. You are only allowed two packets per journey, and the bags are meticulously inspected by the Romanian customs. But if you make five journeys a day, that's €20 profit. Five times a week, that's €100. And €400 a month doubles your state pension. You'd almost be foolish not to.

My colleagues Mircea and Tim are waiting at La Vama café on the Romanian side. They stand out from the thick-set men, all of military age, sitting at tables watching the bridge. Everyone is acutely aware of everyone else, and of their motives.

The first of the bridge-watchers we meet is Mitică Laver. He and his wife Adriana are volunteers, helping women, children and the elderly cross in both directions. Mitică was working as a bouncer in Portugal

when the war started, and came home to help. He has a thin, determined face, a fox of the border woodlands. Watching him in action,
approaching weary women and children on station platforms, carrying
bags for refugees, I'm struck by his compassion. He's the ferryman
across the river Styx, the unexpected angel of the border zone. Each
morning they meet the 09.12 overnight sleeper from Bucharest in their
black van, and offer free rides across the bridge to the railway station in
Solotvyno.

There they meet the midday sleeper from Kyiv, and drive those who
have permission to cross over the bridge to Sighet (as everyone calls it
here) to the railway station. They reckon they helped more than 30,000
Ukrainians in just the first twelve months of the war. They also channel
grateful clients to the 'love hotels' on either side of the border. Ukrainian
soldiers serving on the front, trusted by the military not to abscond, are
allowed to spend their rare spells of leave here. Their wives or girlfriends
(sometimes their children), now living in exile in the West, meet them
at hotels with names like Viza or Fiesta for romantic weekends in
Solotvyno. Meanwhile, in Sighet, Ukrainian men who had jobs in the
EU before the war and decided not to return to fight wait in similar
establishments for their wives and girlfriends to cross from Ukraine to
Romania to visit them. These dreamy backwaters have become unlikely
sex capitals of eastern Europe.

Mitică and Adriana suggest a quieter place to talk than La Vama, so
we drive a few kilometres to a small lake with pollarded willows close to
the Tisa. We can see Ukraine clearly on the far bank, an Orthodox
church with gilded onion domes sparkles in the May sunlight. The river
is shallow here, one of the favourite places for men to wade across.

A nuthatch hops very close to us, as though listening to our conversation; then it returns to the willow to hop carefully round the trunk.
The birds are very loud. Sometimes we hear a splash, as a fisherman
casts his line far out into the lake.

Adriana was on the river shore one February day when two young
men swam across. As they came up the bank, she saw they were very

agitated, and one plunged back in, as if to swim back. She couldn't understand. Eventually he returned, pulling a body. Paramedics tried to resuscitate him, but it was too late for the boys' fifty-five-year-old uncle.

Mitică shows me videos he shot on his phone of family reunions on the Ukrainian side. A man steps forward to pick up his small child, and swings him round in the air. The child pulls away, back to his mother: his father had been away for so long that he didn't recognise him.

'One of the saddest moments I witnessed was at the beginning of the war,' says Mitică. 'A family came across, a mother with two kids. They wanted to reach Germany, so we arranged transport for them. Just before they left, the older daughter, about nineteen years old, got a phone call. Her father had been killed in combat. She hugged me, sobbing and said, "I no longer have a father."'

He is ambivalent about those who don't want to fight. 'I understand them as well. If you're not ready to go there, you shouldn't go. I have friends who fight on the front line. And they tell me that they keep asking their commanders to stop bringing people who don't want to fight, because it actually harms the war effort more than it helps it.'

Most people you meet in Ukraine fear the war will last a long time, but Mitică bases his optimism on a conversation with a Ukrainian general, whom he drove to Budapest once. 'The war has to finish in 2023,' the general told him. 'Because there are no more resources for this war to continue, on either side.'

We go on patrol with the Romanian border police. Maria is twenty-two, a star of the police due both to her beauty and her excellent Ukrainian. Her mother is a Ukrainian Romanian. She never thought the language would come in so handy in her chosen profession. She's often the first person Ukrainian lads meet when they swim across. They must think they've met an angel. 'I try my best to help them and take care of them, and reassure them we won't send them back to Ukraine. They're often very scared.'

She opens the boot of her jeep to show us a cardboard box full of towels, grey and pink blankets, even a pair of jeans and a couple of sweatshirts, all supplied by Mitică and Adriana. The men get dry clothes, food and a trip to the hospital if they need medical help.

At La Vama café I meet Gheorghe, a big guy in a baseball cap with a teddy-bear face, who swam the river in June 2022. He squints awkwardly when he gets to the most painful bits of his story.

The smuggler arranged for one of the border guards to lean a ladder against the razor fence which separates the main road from the river. It was early morning in summer, the moment the first light appears on the horizon. I jumped down the other side, pushed through the under-growth, then got into the river. To my surprise, it wasn't too deep, just up to my chest, so I didn't even have to swim. I just waded across.

Then he heard shots. A patrol of the Ukrainian border guard had spotted him. 'But I wasn't scared. When you've spent time at the front, you know the difference between bullets fired in the air and those aimed at you.'

They shouted insults at him. He couldn't have cared less. But an old leg wound had reopened when he jumped down from the fence and crossed the river. His shoes were full of blood and he needed medical help. I ask him what he would have shouted back, if he could.

If people accuse me of being a traitor I would tell them to go to the front line themselves and find out what it's like. Then come back and call me a traitor. I don't think it's fair, because I've been at the front and I have fought.

His first experience of the front, near Bakhmut, was in April 2022. His hands shake as he shows me photos on his phone of his former comrades. 'All dead now, except this one' – a woman in camouflage uniform.

We were taken straight to the front line. It was terrifying. We had to dig trenches. Almost as soon as we arrived, we came under rocket fire. It was as if the Russians were waiting for us.

I thought I wouldn't make it through that first night. I can remember shooting and a lot of screaming. There was chaos. We had fifty-seven injured and twenty-seven dead. One day about two months later I got injured. There was incoming artillery fire and I jumped into a ditch and broke three of my fingers. As I dived into the trench, head first, one of my legs was hit by shrapnel.

I was there [at the front] for several weeks and we started to learn how to fight. We sort of got the hang of it. At one point we pushed the Russians back 6km, and that felt really good. Our commanders were really proud of us. It even got into the news. But no one knows how much we had to endure.

His wife talked him into deserting. He had two children at home. 'Do you want them to grow up orphans?' she begged him. His fellow soldiers helped him escape the front. He told his commander he had to go back to fetch his phone charger from the building they had slept in the previous night. From there he started running, westwards, all night, parallel to the road through the undergrowth.

At a certain distance from the front, he met his 'helpers' at a pre-arranged spot. For €7,000, they transported him 1,500km across Ukraine to the far western border, dodging or paying off patrols all the way.

In Sighet I find more Ukrainians willing to talk, provided I change their names. I meet two young men in their late twenties – let's call them Liviu and Ion – in the café of the Casa Iurca.

At the start of the war, Ion volunteered for the local territorial defence unit. He got his call-up papers days before his first child was born. He was with his wife for the birth of their son, then went home, where a second call-up notice was waiting. Instead, he swam the river, while his wife pushed his baby boy across the bridge, almost overhead.

He plays me a video of his favourite Ukrainian band, Skofka, a song called 'Chuty himn' – 'Hear the Anthem'. The words are patriotic, 'about fighting and getting back our territories in Bucha, Irpin . . . even

Crimea. It's a song about going back home one day and rebuilding our house.'

'So you still feel patriotic, despite refusing to fight?' I ask.

'You cannot call someone patriotic when they take this sort of decision,' Liviu answers for him. 'When you're a normal person, minding your own business, they take you to the army without any right to say no. I don't feel attracted to that idea of patriotism.'

◉ ◉ ◉

The house where the Jewish writer and Holocaust survivor Elie Wiesel was born is next to the Casa Iurca inn. It's a pale-blue building with white-framed windows and doors in a courtyard of its own, a museum dedicated to Wiesel and the Jewish community, who were deported and perished in the death camps.

'Sighet was a typical shtetl, a sanctuary for Jews . . . since 1640, when refugees began arriving from Ukraine, fleeing the pogroms and persecutions of the reign of Bogdan Khmelnitski,' Wiesel wrote in his memoirs. His father Shlomo, a devout man, ran a grocery store. There's a photograph of him as a young man, with big ears, moustache, short beard and beautiful eyes.

Photographs recall Jewish life in and around Sighet. Many of the Jews of Sighet, and in the foothills of the Carpathians on the far side of the Tisa, were poor. There are prints of men with scythes, their long beards like the great hayricks dotting the steep meadows behind them.

An elderly couple with pitchforks stand in front of another hayrick. A metal worker in Sighet looks up from his anvil, hammer resting in his right hand, a crowd of young, laughing faces all around him. In Herbert J. Seligmann's photograph of market day in Sighet in September 1937, an old man sits on the cobblestones, his potatoes and onions spread out on sackcloth beside him.

Wiesel hardly knew his father. The only day they spent together was the Sabbath, which began on Friday afternoon. He would walk holding hands with his father past the prison. When someone approached to talk to his father, he would let go of the boy's hand. Even in old age, Elie still remembered that terrible sense of loss:

> I admired him, feared him, loved him intensely. He, in turn, genu-
> inely loved all people – the weak, the needy, even the madmen. He
> enjoyed listening to them as they laughed, sang, wept and chattered
> with birds they alone could see. Beggars were drawn to him and he
> never failed to invite them to share our Shabbat meal.

Elie was very close to his mother, Sarah Feig, from the village of Bychkiv, just upriver, in what is now Ukraine. In August 1940, just before Elie's twelfth birthday, Sighet was transferred back to Hungary under the Second Vienna Award. Hungarian units appeared in the town riding bicycles, and the local population, largely Jewish and Hungarian, turned out on the streets to cheer them. 'My mother, too, was pleased with our change of nationality. For her it was a kind of return to her childhood, for which thanks were due to God.'

News of the war filtered into Sighet through the Hungarian and Yiddish-language newspapers, especially of the plight of the Jews in Germany. But it seemed far away.

'In spite of everything we knew about Nazi Germany, we had an inexplicable confidence in German culture and humanism. We kept telling ourselves that this was, after all, a civilised people, that we must not give credence to exaggerated rumours about its army's behaviour.' During the First World War, the German army had rescued Jews who had been beaten and humiliated under Russian Cossack occupation.

Operation Barbarossa began in June 1941. The German army, with its Romanian and Hungarian allies, invaded the Soviet Union in the largest land offensive in history, with an army of 3.8 million along a front stretching 2,900km. Hitler's aim was to occupy the west of the USSR, in order to repopulate it with Germans. The great fields of Ukraine, the oil of the Caucasus and the slave labour of non-Aryan people were needed for the German war effort. This was the main thrust of Hitler's scheme to create *Lebensraum* – 'living space' for the Germans – and to exterminate the Slavic peoples.

Jews who could not prove their Hungarian citizenship were expelled across the mountains to Polish Galicia. Some 23,000 were massacred in Kolomyya, Stanislav and Kamenets-Podolski at the end of August. Just one, Moshe, made it back to Sighet, half-crazed with the horror of what he had witnessed. But no one believed his tales of Jews forced to dig their own mass graves.

In March 1944, the Nazis occupied Hungary, impatient with the implementation of the 'Final Solution'. German soldiers appeared in the town, and within days new decrees were issued. All stores belonging to Jews, including that of the Wiesels, were closed down. Jews were forbidden to go out on the street, apart from a short period each afternoon. Yellow stars had to be worn – the women made them themselves, sharing yellow cloth. House searches by German soldiers or Hungarian gendarmes began. Jewellery, silver, gold and foreign currency had to be handed over. They hid what they could.

> The truth is, some Jews in Sighet could have escaped the ghetto. It was a mild spring, and they had only to flee to the mountains until the ordeal was over. Maria – our old housekeeper, wonderful Maria who had worked for us since I was born – begged us to follow her to her home. She offered us her cabin in a remote hamlet. But we didn't leave, no matter how much this 'simple Christian peasant woman' begged us. No one, except Moshe, had any idea of the coming nightmare.

The Jews of Sighet were shut into a ghetto. On 15 May 1944, they were marched to the railway station by the gendarmes. From 16 to 22 May, they were packed into four transports. Elie Wiesel and his family were in the last. With bureaucratic efficiency, the numbers for the four days were recorded exactly – 12,849. You can see them in the blurred black-and-white prints on the walls of Elie Wiesel's house.

Throughout the terrible days of the journey to Auschwitz, Elie's mother exhorted the family – Elie and his three sisters, Shlomo and her own mother – to stay together at all costs. On the platform they were separated. Elie was fifteen, but said he was eighteen. He and his father were sent to slave labour. His mother, sisters and grandmother were marched straight to the gas chambers.

Most of the tracks at Sighet station are overgrown with grass. I climb up into a rusting locomotive. There are small trees growing on the footplate. Among the grasses and the scarlet poppies, I gaze along the tracks which took the Jews of Sighet to their doom.

The marketplace feels strangely modern, with its stalls and fresh summer vegetables. Fat, pale oxheart tomatoes. Piles of porcini mushrooms in all shapes and sizes, fresh from the forest.

◎ ◎ ◎

After the war, the calvary of Sighet continued. In the nineteenth-century prison in the heart of the city, prominent 'enemies' of the communist regime were imprisoned in atrocious conditions. Politicians, priests and ordinary folk who attracted the regime's wrath were incarcerated in a jailhouse designed by the architect Gyula Wagner, who has no fewer than twenty-five prisons to his name.

Romania joined the Second World War on the German side, then switched to the Allies in August 1944. The young King Michael I, in the course of a private audience, ordered palace guards to arrest Romania's de-facto leader, Marshal Antonescu. I interviewed Michael on his ninetieth birthday, beside Lake Geneva in Switzerland. 'I had Antonescu locked up upstairs,' he grinned, 'in the little room where my father kept his stamp collection.'

Romania ended the war on the victorious side, but under Red Army occupation. By changing sides – albeit at the last minute, when it was clear Germany was losing the war – Romania got back all of Transylvania. Winning back the mountains was the central plank of Hungary's foreign policy from 1920 to 1940. In 1945, she lost them all again.

From 1946, the Communist Party in Romania, with the Red Army at its back, moved swiftly to ban other parties and to arrest or force their opponents into exile.

King Michael held on for a while, and against the advice of Winston Churchill, returned home from the royal wedding in London in 1947. In December, he was forced to abdicate. The communists put him on a train to Switzerland with his mother, his Alsatian dog and a few possessions from his palace at Sinaia, high in the Carpathians, near Braşov. The Romanian soldiers who normally formed an honour guard to greet him had been ordered by Romania's new rulers to turn their backs on him as he boarded the train. As he passed them, he told me, he saw their cheeks were streaming with tears.

The Romanian Communist Party, led by Gheorghe Gheorghiu-Dej and then Nicolae Ceauşescu, governed Romania with an iron fist for forty-one years. More than 200 members of the non-communist elite, leaders of political parties, bishops and priests were imprisoned in Sighet between May 1950 and July 1955. The town was chosen because the Soviet Union was just the other side of the Tisa river. In the event of an uprising against communist rule, the prize inmates could be swiftly spirited over the border into the bosom of an ideologically friendly Soviet Union.

In the main hall of the prison, there's a mosaic of photographs of the former inmates. What facial expression should one pull when one is photographed by one's captors? Some look hunted, others defiant; but the most striking are those who stare straight through the lens into the future. One day, they knew, someone living in freedom might gaze on their imprisoned faces, and their eyes might meet.

The prison is on three floors. Iuliu Maniu, prime minister of Romania from 1928 to 1930, was one of fifty-five men to die here, aged eighty. His party was abolished by the communists in 1947. His body was thrown into a mass grave in the courtyard of the prison, and the clerk wrote 'unemployed' in the box marked 'profession'.

One of the bigger cells has an exhibition on the anti-communist partisans who held out in the southern Carpathians for a full decade, before they were betrayed and executed in 1958. In another cell is an exhibit dedicated to the uprising in Braşov in 1987, the precursor of the 1989 revolution.

I'm about to leave when Ana Blandiana, one of the founders of the museum arrives.

We talk in the 'space for prayer and recollection', a womb-like structure in the yard with a circular wheel in the middle, in which people light thin, beeswax candles. At the age of seventy-nine, she's a striking figure with a powerful voice.

'The revolution did not achieve so much,' she says, as a group of Italian Catholics try to interrupt to invite her to Bolzano, 'but it did achieve freedom.' Her fellow Romanians have not known what to do with that freedom, she says, except to leave the country. The whole of Romania under communism resembled the prison in Sighet. Once the doors were opened, the inmates left.

◎ ◎ ◎

Highway 18 edges east past the headquarters of the border police, then curves away from the border, round the edge of the Maramureş massif. Vişeu de Sus is in the valley between the Maramureş and Rodna mountains. The narrow-gauge steam railway, the *Mocăniţă*, was built to haul timber out of the forests of Maramureş, but its main cargo now is tourists. The locomotive is black with gold lines and red plates, built in Reşiţa, maximum speed 30km per hour. As it builds up steam, smoke billows from its funnel, refracted magnificently by the sunlight streaming through the tall spruce trees beside the track. Steam hisses or drifts from four other places along the locomotive.

It's a six-hour round trip, 20km along the shore of the River Vaser as far as Paltin, and back. The carriages are painted green and carry the unlikely destination plate Wengen–Lauterbrunnen, their home villages in the Black Forest in Germany. The driver invites me up onto the footplate. Denis's forearms are stained black by the coal dust. His father drove the same locomotive, and his grandfather before him.

The engine runs on great timber logs, thrown into the flaming furnace as we chug along beside the river, bearing a bouquet of grey-black smoke through the branches of the trees. Each time Denis opens the round plate to throw in more fuel, we're struck by a blast of heat.

I pull a lever to sound the glorious horn, echoing through the mountains. You can probably hear it in Ukraine. There are so many levers to pull and adjust, it feels like being inside the engine of a giant motorbike. Denis's mother and father are working in Germany, his sister in France.

The railway was built in the 1930s, to replace the practice of floating the logs downriver. The idea of such narrow-gauge railways was simple – the tracks were built on tight curves to follow the river, the locomotives hauled the empty timber wagons uphill, and then heavily loaded trains rolled downhill to the sawmills.

The railway was badly damaged by the retreating German and Hungarian armies in the autumn of 1944, restored and extended into other valleys during the 1950s, then rescued from oblivion by a Swiss railway enthusiast in the late 1980s.

At Paltin, the train stops for lunch. Denis broaches a Timişoreana, a rather palatable beer. We wander down to the stony shore of the Vaser, to sit in the shade beneath the willows. Then we huff and puff all the way back to Vişeu.

◎ ◎ ◎

Radu and Irina run a restaurant in an orchard next to the campsite where I pitch my tent in Breb, a pretty village at the foot of the Gutâi hills. Breb was made famous by the British author William Blacker, who first came here in 1990. The impact of the visitors and their money is visible everywhere, but the beauty which drew them has not quite faded.

Radu and Irina arrived in 2014, and began by buying sheep and restoring the traditional wooden houses. Irina is fresh-faced, with blue eyes and a polka-dot dress. She's a champion of sustainable tourism, concerned about the large-scale industry that delivers big groups to Breb by the coach load. The jagged teeth of the Creasta Cocoşului (Rooster's Crest) overlook their garden, 1,200 metres above sea level.

'The tourists come and eat, but they don't interact with the locals at all,' she says. She's happy, on the other hand, with all the young people who have been drawn to live here and contribute to the community. But the locals are turning away from the fields, to concentrate on offering tourist accommodation. In her eight years here, many have sold their land and their animals. Although there are still beekeepers in the village, and carpenters who know how to carve the high, ornate gates. People still think of this region as 'poor', Irina adds. They think the outsiders come just for 'the food and the air'. They don't realise how rich they are, how valuable this place is in the grand scheme of things.

The wooden houses are made from pine in Breb, oak in other places. There are even villages where they are made of larch, because of the plentiful local supply. The pine keeps the warmth better, while oak houses stay cooler in summer. The churches are always made from oak.

The wood is cut at the time of the full moon, then left to season. The buildings are assembled on the ground, each plank and timber piece carefully numbered. Then the construction work can begin, the larger pieces hauled by a tractor.

The people of Breb work abroad, then bring the money home and build large, ugly houses that stand out from the rest. 'No one ever told them how to spend money in a sensible way, to enjoy themselves on holiday, or on their own health.'

We watch a hawk hovering over the meadow, selecting its prey in the fresh summer morning. Breb has remained a living place and community, despite all the changes, Irina says. She contrasts it to some other 'tourist centres' in Romania, which by now revolve just around the visitors. In Breb, some newcomers have learnt carpentry and other skills from the locals.

Thursdays are market days in nearby Ocna Şugatag – Holy Thursdays, the locals joke, because no one works that day. It's haymaking time, and the tall hayricks are multiplying in the meadows, as the long grass falls to the sharp blade. Winter is the time to chatter to visitors in these parts, not summer.

I find Marioara at home tending her bees in two long rows of multi-coloured hives, near the edge of Breb. She casually offers her hands to her bees to crawl over. Then she pours the honey into big white plastic buckets, and I buy a couple of jars in her small shop. The honey is delicious.

That night there's a powerful storm in the valley, and I have to get up several times to reattach the guy-ropes of our small tent.

At sixteen, Valentin speaks rather good English, learnt from the computer game Minecraft. 'Family friendly,' he announces, apropos of nothing in particular. I guess it must be a version of Minecraft without story lines that might upset the little ones. He shows us round the Church of the Archangels in Breb. The interior is bedecked with patterned woollen rugs. We want to take off our shoes, as one would in a mosque, but he says it's not necessary. Christians, I suppose, mostly pray with their boots on.

When we drive up the pass from Budeşti towards Cavnic, there are Gypsies by the roadside, selling big buckets of blueberries and wild raspberries. We bargain for a bucketload. A small man with bright blue eyes, straw-coloured hair and terrible teeth introduces himself as Gábor Révész – literally 'Gabriel Ferryman'. He's been all over, he says, and lists towns in eastern and western Hungary – Debrecen, Tatabánya. He worked in the gold mines for ten years and filled his pockets with precious stones, which he sold in Hungary. A big woman with a round, generous face stands to one side in billowing skirts, a small boy running rings around her. Beyond them is a cluster of Gypsies with nothing to sell, but hoping to cadge a lift to Cavnic.

The town's main claim to glory, apart from the mines, was the defeat of the Tatars here in 1717, on their last foray into eastern Europe.

There's an obelisk at the top of the town to commemorate the event, next to a spring where we fill our bottles with ice-cold water and try to rinse the blueberry stains from our hands.

Back at the campsite, the sun sets beyond the Rooster's Crest. I drink a small blackcurrant *pálinka* and fall into a fruitful sleep.

CHAPTER 13

TRAVELS IN AN ARMCHAIR

The Rosebay Mountains. Streets strewn with river rocks, wood-gabled villages kilometres long, homes without chimneys, centuries-old wooden churches – onion-shaped – and paschal candles half a millennium old.

Miloslav Nevrlý, *Carpathian Games*

I sit in my pale beechwood armchair in Budapest, dreaming of Transylvania. The chair is a stylised 'U' shape, fitted with a beige cushion, and it has a pleasant bounce to it. The wooden structure, glued beechwood veneers, is easy on the eye and the fingers. My back feels well supported. It's a light, modern armchair, popular and affordable.

The design, from the Swedish IKEA company, is called 'Pöang' and was first introduced in 1977 in tubular steel, then redesigned in beechwood in 1992. Designer Noboru Nakamura wrote that his intention was

to combine the Scandinavian and Japanese furniture traditions. It also felt important to make the most of the qualities of layer-glued wood. That's why the frame is a little springy, which makes the armchair even more comfortable.

Nakamura passed away in April 2023. May he rest in peace, in one of his own chairs, in designer heaven.

The only problem with the chair is that it's made from beech trees, and IKEA is not always as careful as it should be about where that beech is sourced. Some is from forests in the Carpathians, in Romania and Ukraine, which are being illegally logged – or, at best, over-logged. IKEA is the largest private forest owner in Romania, with 50,000 hectares bought since 2014 by an IKEA subsidiary, Ingka Holdings. On paper, it's a model company, a champion of sustainable forestry. Its publicity material purrs with good intentions. The problem, environmentalists from Agent Green and Greenpeace claim, lies in the execution. In the traditionally corrupt world of Romanian and Ukrainian forestry, it's easy to turn a blind eye to the way things have always been done at a local level.

I drive with Gabi Paun and Ion Barbu through the grubby town of Borşa, beside the Maramureş mountains towards the Prislop pass, its head buried in the late August clouds. Then on towards Cârlibaba, and left onto a mountain road to follow the Ţibău river upstream.

Spruce trees grow thick on the slopes above the track. We stop for a break and I walk barefoot on the hard muddy shore of the Ţibău, a small but muscular river. Beside my footprints are the perfect paw marks of a fox. Small bees busy themselves among the white dog roses and purple field scabious, known in German as *Wald-Witwenblume* – 'forest widow's flower'.

We see signs of cutting, and park the car discreetly. It's Saturday, so most likely the foresters are home with their families. But Gabi and his colleagues have had too many brushes with angry men over the years to take our safety lightly. Usually there's just a shouted exchange, some verbal threats and a bit of jostling. But he was beaten up badly once. Forestry guards have fared worse: from 2010 to 2020 there were over 600 physical attacks on members of the Forestry Guard when they tried to intervene to stop illegal logging. Six guards, including one in Maramureş county, were killed.

White and grey clouds clog the sky over Ukraine in the distance, and we feel the first few drops of rain on our faces.

This plot, on a small hill surrounded by clearcut forest, is owned by IKEA. According to Google Maps, it should be a thickly forested hilltop, but it has all been cut. Among the tree stumps, the branches are scattered. The surface of the puddles glistens with waste fuel. At the foot of one of the few trees left standing, we pick wild raspberries. It would be unfair to say the hilltop has been clearcut: under Romanian law, that phrase can only apply to areas over 3 hectares in size. Instead, they call this 'progressive logging'. That means you take out most of the trees one day, then come back for the others months or years later. The environmentalists argue that this amounts to the same thing: all the trees are cut. The foresters argue that it gives 'natural regeneration' a chance.

From the point where the Ursului river flows into the Țibău, right up to the Ukrainian border, almost all the trees have been cut in earlier years. This particular destruction was carried out by Romsilva, the state forestry company, on state-owned land.

In August 2021, Agent Green issued a thirty-page report, with five documented local studies, accusing the Swedish furniture giant of 'actively seeking out old growth forests on its property for logging before any strict protection can be implemented by law'. Agent Green also wrote that after 120 years, trees begin to lose their commercial value, just as their biodiversity value increases.

Beechwood, the report points out, changes colour slowly as it gets older. The prime time to cut the trees would be between 80 and 100 years, when the wood still has that pale hue, which is so attractive in my armchair. After a hundred years, the wood starts to turn darker shades of red and pink, and it becomes knottier – so of less commercial value. The battle is now for these older trees (a beech can live to around 300 years). Year by year the ancient beeches are becoming more valuable citizens of the forest, home to a multitude of bugs and birds and fungi.

I ask IKEA to reply to the Agent Green report, and to our findings in the Ţibău valley. In his reply, Andriy Hrytsyuk, Global Forest Land Investments Manager at Ingka Investments, writes:

> Our main priority as forest owners is to ensure responsible forest management of all our properties that protects the forest, environment, and biodiversity for many generations to come.
>
> Our objective is to minimize our impact on the environment . . . and we are actively committed to improving our operations and exploring better ways of working.
>
> We do this by taking a long-term approach that respects the law, sets a high standard, and remains transparent and open to suggestions as to how we can do better.

The company does not own any virgin or quasi-virgin forests in Romania, he adds. Much of the debate about forestry in the Carpathians hinges on this question of definitions. What value to assign a stand of beech or fir that is more than 150 years old? Or 130 years old? At what point should it be tagged 'worthy of protection'? Another major issue, as Erik Baláž pointed out in the High Tatras in Slovakia, is for large enough continuous areas of wilderness to be left alone, to allow the relationships of the wild, from fungi to large predators, to take place without human interference.

There are still primeval forests in the Carpathians that have never known human intervention, and many are strictly protected. There are many more 'old-growth' forests which have hardly been disturbed by man over the centuries simply because they were too remote, or on such steep hillsides that it was impossible to extract wood with the technology available. There are also 'buffer zones' around the virgin forest in national parks. Advances in satellite mapping and in cutting and transporting technology mean the forestry companies can drive their roads deeper and deeper into the forests, and extract timber that was out of reach until now.

Under communism, the forests were nationalised. Since 1990, they have gradually been re-privatised – a long and controversial process. By 2023, 65 per cent of forests in Romania were in private hands. Many private owners want to make money quickly, so the pressure on the forests is growing all the time.

Activists are using technology to try to hold the foresters to account. An app for mobile phones, Forest Inspector, has been downloaded by more than 80,000 people: it traces any cargo of timber from chainsaw to sawmill. If you are driving down a mountain road behind a timber lorry – and have a good enough signal – you just type in the registration number of the truck to find out the size, origin and destination of its load. The system is not watertight, however, and the loggers have found ways of continuing to strip the forests of trees at a faster rate and in greater quantities than they admit.

In Cârlibaba we talk to a family – mother, father and teenage daughter – chopping wood at the roadside. 'Since the big companies started exploiting forests in this area,' replies Vasile Potiuc, 'we have to wait months for firewood. When it finally arrives it is of low quality and costs nearly twice as much as a year ago. They tell us the forest is exhausted. But we know there is wood for the big companies, just none for us, the locals.' His wife keeps chopping wood, while their daughter dances up and down the ladder, her hair tied up in a bun. The stack is nearly ready. 'I spend most of my money on medicines. It's frustrating because I know the wood is rotting away in the forest, where the logging companies leave it, yet we can't get our hands on it.' The family depends on the money sent home by his older sons in England.

'Is the corruption foreign, or homegrown?' I ask. '*Noastra*' – ours – he laughs bitterly. 'Yes, there are GPS trackers on the trucks, and sometimes inspectors go into the forest and check; but the theft is so big, I don't think they will ever be stopped.'

◎ ◎ ◎

Radu works for Romsilva, the Romanian state forestry company. It was established in 1990 and is responsible for both exploiting and protecting the forests of Romania. Though only 35 per cent are still in state hands, Romsilva is responsible for 65 per cent of forests, since many of the privately owned lands are in national parks, which Romsilva administers.

Cutting and protecting is not an easy combination of roles. Two thirds of Europe's remaining old-growth forests are in Romania, so forestry is a major source of income locally and nationally. Three million Romanians depend on firewood to heat their homes. Environmentalists accuse Romsilva of being at the beck and call of local and national politicians, into whose coffers the proceeds from the illegal sale of wood flows. Managers in the wood industry retort that they have a working business, carefully regulated by law, and that timber is a national resource which they have a right and a duty to exploit.

The denuded Carpathian slopes shed their topsoil into the streams and rivers. The silt is trapped in the storage lakes of the dams – 178 large dams in Romania alone.

The difference in ownership between Ukraine and Romania means that the forests in Romania and Ukraine are milked in different ways. In Ukraine, forest inspectors are paid to say that healthy forests are sick and may therefore be cut; in Romania, a wide range of other tricks is deployed.

'The system itself is bad,' says Radu. 'Our income depends on the volume of wood sold, so the incentive is to cut as much as possible.' If the salaries of Romsilva employees could be decoupled from the amount of wood they deliver, there would be less illegal or over-logging, he suggests.

From this region, the trees are loaded on trucks and sent to the giant sawmill at Sebeş in Alba county, 280km south.

Wood cutting is divided into ten-year management plans. In the 2009–19 plan, 100,000 to 150,000 cubic metres of timber were cut in Maramureş county, about a twentieth of the wood cut in the whole of Romania. The ten-year plan for 2019–29 contains a similar allocation.

According to the Romanian Forestry Inventory, 18 million cubic metres of timber were cut legally each year in the country from 2014 to 2017, but a further 18 million cubic metres were cut illegally each year. In 2020, the European Commission started a case against Romania at the European Court of Justice for failing to tackle illegal logging.

Successive Romanian governments have tried to find a technical solution. The Forest Inspector app has reduced illegal logging, but is not perfect, Radu explains:

Let's say Schweighofer [a major Austrian company, renamed H.S. Timber] buys timber from 1,000 suppliers here in Maramureș. My boss sends me to mark a plot of forest for felling, let's call it plot 22B. A few years ago, I just wrote a paper. Now we have to take photos of the back, front and both sides of the trucks. Then we can track the whole journey by GPS.

He cites Poienile de sub Munte as an example of illegal logging. In the summer of 2021, I visited Poienile. The spruce and deciduous trees on the slopes of the Ruscova river are lush, but the trees thin out dramatically near the source, close to the Ukrainian border. Satellite photographs show just the zig-zag forestry roads – all the trees have gone.

Driving through Repedea on a Saturday morning, we met a wedding procession walking the other way. A solemn bride and groom led the guests along a pot-holed, dusty road. She wore a bridal dress of angelic white, and was followed by bridesmaids in lace-fringed linen blouses and traditional peasant skirts, brown with horizontal stripes, their pretty, pony-tailed heads bowed to concentrate on keeping the bride's dress out of the dust. They were followed by several dozen villagers, the women in headscarves and the men in white shirts. There was hardly a smile on their faces. Perhaps such solemnity is expected.

In Vişeu de Sus, I meet Beatrix Szabó, environmental officer for the Maramureş National Park. She, too, works for Romsilva. Originally a biologist, she was first drawn to the Rodna mountains to study alpine marmots. The legislation is improving, she says, a little defensively, citing successful efforts to protect some species from hunting. On the downside, apart from the deforestation, she lists all the gravel extraction from riverbeds to fuel the booming construction industry. The excuse usually given is 'to enhance flood protection'; but the digging makes the flooding worse, as the flood plain of the rivers, which used to absorb the flood waters, is narrowed into a single channel. One of Bea's major successes was to stop the damming of the Ruscova river for a micro-hydroelectric project. Much later, all new micro-hydropower projects were banned at county level. Her story shows the ambivalent role played by Romsilva: sometimes the culprit, sometimes managing to ameliorate or repair the worst damage to the mountain landscape.

According to Bea, the private owners do the most damage to the forests. 'I'm the boss, I own it, and I don't care what you say, I'll do exactly what I want with my trees,' is the prevailing mood.

◎ ◎ ◎

'Far up the 2,300 metre high ridges of the Rodna Alps, beauty awaits,' wrote Miloslav Nevrlý. 'Realms of grass, verdant mountain crests and millions of gentle-red rhododendrons. Here, the weather shifts from storms to mist to sunshine, sometimes twice a day.'

We set out to walk through the Rodna mountains, which the Hungarians call the 'Radnai Snowies'. They call all the mountain ranges 'snowies' – in memory of the time when we still had snow.

We park the car near the Şetref pass and strike out on foot up a well-marked track. It's still early morning, at 817 metres. We pass a young man just getting out of his tent, bleary-eyed. The track leads to the tallest mountain in the range, Pietrosu.

The route is well marked, and soon we pass through a pine forest on a broad track. The bark on a tall fir has been assiduously rubbed away, and the mass of hairs protruding from the pine resin prove that this is a bear tree. We come out on an open meadow with a spring in the middle, and bump into the man we saw emerging from his tent that morning.

He came here, he explains, to clear his lungs after several years of working in a fertiliser plant near Constanța, on the Romanian Black Sea coast. He's trying to summon up the courage to quit his job.

Just as the path climbs steeply again, we look down into a valley from which nearly all the trees have gone.

A week later, on a chilly August evening, our rented, four-wheel-drive car reaches the end of a forest road in the hills above Borşa. We sleep in the most beautiful, Hansel and Gretel cottage in the mountains: low-beamed ceilings, a green-painted wooden planked floor, traditional wooden furniture and pretty rugs on the white walls. Icons, too, painted on glass in the naive, Maramureş style. The Virgin Mary surrounded by flowers and thatched roofs. And Christ himself, on the Cross, in what looks like a forest with trees covered in white blossoms. Jesus of Nazareth has been transposed from a landscape of olives and palms to one of beech and fir, and the ruddy glow on Mary's cheeks can only be from the good mountain air.

There is no food in the house, and we have forgotten to bring our own. We settle down to a humble dinner of biscuits and strong liquor. I sleep in a green wooden bed with a red-and-white woollen blanket on a rather lumpy mattress, and wake before dawn to walk barefoot outside, in grass awash with dew. Low mist is rising in the mountainous valley to the east, in a sky streaked pink and yellow. In the foreground, hayricks float like small houses in the meadow, cut free from the ground by the mist. What I thought were roofs in the icons were probably hayricks: the roof of our house is made of dark wooden slats. Scarlet geraniums bloom along the little porch, beside grass-green shutters.

CHAPTER 14

HEARTS LIKE FLAGS

The predominant population, composed of Romanians and Ruthenians, is fearless at times of danger, unforgiving to accusations, and inclined towards changing the government.

Baron Splény de Miháldy (1734–1818), Austro-Hungarian governor of Bukovina (1774–78)

Bukovina nestled up at the top right-hand corner of the old Austro-Hungarian empire, alongside Transylvania and Galicia. Bukovina was the smallest. Transylvania was the torso, holding the other two up. Galicia was long and thin, perched on the shoulders of the other two like a sack of potatoes.

The traditional lands of Moldavia were three – Upper, Lower and Bessarabia. There is still a Basarab railway station in Bucharest.

In 1774, Austria contributed to the Russian victory in the Russo-Turkish war and was given Bukovina in gratitude. An outlying province of the Austrian empire, it was rich in trees, minerals and monasteries, but never quite qualified as a kingdom. In the Assembly Hall of the Palace of Justice in Vienna, you can still see the red-and-blue coat of arms of Bukovina, with the head of a bull in the centre.

Bukovina has historically had two capitals: Chernivtsi to the north in what is now Ukraine, and Suceava to the south in Romania. Chernivtsi and northern Bukovina were swallowed up by the Soviet Union after the Second World War.

I take the train to Suceava from Vatra Dornei, a pretty town at the confluence of the Dorna and Bistriţa rivers, once famous for its healing waters and the magnificent casino built there in the 1890s. Emperor Franz Josef and Archduke Ferdinand attended its opening in 1898. Nicolae and Elena Ceauşescu were due to spend New Year's Eve there in 1989, but were inconveniently overthrown and executed a few days earlier. Since then, the casino has fallen into ruin, torn between the Orthodox Archdiocese of Suceava and the city council. There are plans to turn it into a museum.

The railway follows the Bistriţa river, then branches east at Bistriţa Aurie towards Suceava. Sitting opposite me is Per, a gardener from Sweden. He's backpacking through the Carpathians and has just crossed the Low Tatras in Slovakia from Bratislava to Košice. The Covid pandemic is still in full swing. Some people wear masks on the train, but most don't bother. There's a healthy scepticism towards the state here, a legacy of communism. A widespread doubt that the government of the day, or its experts, are genuinely interested in the welfare of the citizens – in stark contrast to the trusting peoples of western Europe.

Per worked in a care home in Sweden as a volunteer nurse. He's a great admirer of Anders Tegnell, the Swedish physician and chief epidemiologist, who opposed lockdowns.

At some point in our conversation, a girl in our compartment asks if anyone would like a cup of tea. It's an old-fashioned train, Russian style, with a samovar at the end of the carriage. She even has her own supply of teabags. Soon I'm sipping spiced Indian chai, chugging down the northern slopes of the eastern Carpathians into Bukovina. As we leave the peaks behind, the train seems to speed up through the apple orchards towards Suceava.

At the station, I pick up a hire car, then set out immediately back westwards, towards the sun, setting scarlet over the dark hills of Bukovina.

As one approaches from the east, the town of Gura Humorului stands like a sentinel at the entrance to the Carpathians. The river Moldova seems rather muddy and shallow here. The town has a strong German and Jewish tradition, and is the birthplace of the Ukrainian feminist writer Olha Kobylianska, the lover of Lesya Ukrainka, whose poem the young lad recited to me in Uzhhorod, under the ash tree.

Lucy Glaser is one of the few Germans left in town. She runs a guesthouse set back from the main road, with a garden full of ducks running free, delighting in the early summer apples. I wander down the main street for supper in the Atipic restaurant on the river shore. It's a welcoming place, with a young crowd, excellent cream of mushroom soup and a respectable pizza, washed down with a crisp local wine. Bukovina is a place set apart from the rest of the country, and proud of it. 'If you come from Oltenia,' Lucy tells me the next day, referring to the province southeast of Bucharest, bordering the Danube, 'you would never mention it, but from Bukovina – definitely.'

From the Middle Ages on, Germans wandered to all corners of Europe, where their skills as miners, farmers and builders were much in demand. Lucy's family, the Glasers, were one of the first twenty-nine families to arrive in Gura. The Habsburg Empress Maria Theresa (1740–80) directed them to wherever they were needed across the empire. The Germans came as settlers – and settle they did, until they were ordered 'home'.

In 1938, Adolf Hitler launched his 'Heim ins Reich' programme. The term *Volksdeutsche* – 'Folk-Germans' or 'ethnic Germans' – was applied to the far-flung German diaspora. The logic of occupying Poland and northern Bohemia (the Sudetenland), and of annexing Austria, was that these were all lapsed German territories that needed to be brought back into the fold. However, all Germans living outside greater Germany – and who were under the thumb of 'inferior' Slavs, Latins and the like – should be brought back to the fatherland, or sent to colonise the newly seized parts of Poland.

At the end of the First World War, 9 million ethnic Germans lived in the redrawn countries of Poland and Romania, and in the new countries of Czechoslovakia and Yugoslavia. Ahead of his war, Hitler needed their elbow grease. For the most part they obeyed, abandoning towns and villages they had helped build and had contributed to for centuries, leaving a gaping hole behind them.

Lucy's family stayed when nearly everyone else left, out of love for the animals. 'My father couldn't bear to leave his horses,' she says.

As a young woman in 1980, she remembers a frantic phone call from her mother – 'please come home, immediately!' Out of the blue, a man claiming to be her long-lost uncle Rudolf had turned up with his daughter Marianne at the family home in Gura.

Rudolf was born in November 1903 in Bori, the German suburb of Gura. By 1923, he was in all kinds of trouble with his family and the local authorities: exactly what kind of trouble, no one can remember; but it was enough to convince Lucy's grandmother, Rosa, Rudolf's mother, that the boy should be sent away to relatives in Canada. He was just one of the 357,803 eastern European migrants allowed into North America in the 1920s.

He set sail from Hamburg – the German equivalent of Liverpool – to join his sisters Maria and Paulina in Burlington, Ontario. He got a job on the Canadian Pacific Railway, went out west to British Columbia on the Pacific seaboard . . . and never came back. He survived the Wall Street Crash of 1929 and the Great Depression that followed, but lost contact with his sisters. The family assumed he must have died in an accident. Grandma Rosa always mentioned him in her prayers at the cemetery in Gura, Lucy says.

Suddenly in 1980 here was a seventy-seven-year-old man claiming to be him, in the kitchen of the same house he'd left fifty-seven years earlier. And he had forgotten his German.

Lucy rushed home from her workplace, and together they began to piece together the story. He had married an Englishwoman in Canada, set up home in Alberta, and Marianne was their only daughter. Sometime in the 1950s, Rudolf had lost his sight. Marianne got a job at a company marketing Romanian Dacia cars, which were imported to Canada as a cheap, reliable vehicle from the 1970s, made in Romania under licence from the French manufacturer Renault.

In the late 1970s, Marianne heard of a new laser technology that might be able to help her father see again. The first operation, on his left eye, was unsuccessful. Then they tried his right eye. 'Total success! After twenty-five years, he could see again! He felt like he had been reborn!'

Rudolf wanted to see his birthplace again, so father and daughter booked plane tickets to Bucharest, then to Suceava. The communist authorities forbade foreigners to stay in private houses, so they booked a hotel in Suceava and hired a car. And there they were.

At first the family was sceptical, suspecting a scam.

But then he recognised everything! The school. His grandmother's house, where some houses used to be, but are no longer standing. Everyone got very emotional. An old box full of photographs was pulled out, and he remembered the names of everyone!

So the brother and sister were reunited. They were there for four days, and each day they lunched with the family – foreigners were not allowed to spend more than three hours at a time with Romanian citizens. They spent a wonderful weekend together, but that was all. They had booked flights back to Canada. Too soon, the miracle was over, and Rudolf and Marianne were gone, promising to stay in touch.

'But the strange thing was,' Lucy told me, pouring another strong coffee in her kitchen, as the Bukovinian rain pattered on the windows, 'that we never heard from them again.'

We wrote to them at the address they gave. No reply. We wrote again and again. No reply. And then, in the age of the internet, I've tried to track Marianne down again, but in vain. Rudolph would be 118 now, of course, so he must have died. Perhaps they were worried that we, in communist Romania, might start asking them for money, or to arrange visas for them for Canada. Or something else happened.

An article on the webpage of the Bukovina Society of the Americas offers a sobering thought:

> After the war some Bukovina Germans returned for a visit to their homeland only to find the experience a depressing one, with their ancestral graves leveled, their former homes dilapidated, and the warmth and familiarity of their Heimat (homeland) only a faint memory. Of the more than 75,000 Germans residing in Bukovina in 1930, fewer than 2,000 live there today.

Lucy drives me through the Bori district to her aunt's house. There are sweet red apples all over the courtyard. Her aunt lives in Bucharest now, but visits when her health allows.

Many people in the street recognise Lucy and say hello, but always in Romanian. She is the last German.

Despite that, she has founded a German Forum, which is well attended, with fifty or sixty regular members. People with family ties come back. She tries to teach German to the children, including her own son and daughter. She sews, or buys, folk costumes from the surrounding area:

> I try to teach German culture to the Romanians. It's hard because some of them enjoy and respect it. But others feel the opposite. They don't like to step out of their Romanian identity. There's something like an invisible wall between us. I never felt rejected by Romanians. I was more Romanian than them! I speak Romanian in their dialect.

· The folk costumes used to be different for each village, but now she uses what she can lay her hands on – blue-and-white chequered table-cloths from Bavaria, and blouses from Poland with flowers on the shoulders. Looking through her Facebook pictures, I have the impression that she's trying to rescue tradition, *any* tradition, from the flood of modernity.

Lucy cooks up a pot of tomato soup and shows me the illustrated book her father read to her as a child, *A School Manual of Natural History*. 'He explained not just what animals look like, but how they behave, and how they act towards humans.'

We thumb through it together. The fox was a favourite, and the rabbit. The kangaroo has been underlined in pencil, in a wobbly, childish hand. We skim though various woodpeckers, the hoopoe and the storks. There was playful, moral instruction while they studied the book together. 'Don't be like the summer birds, singing all summer long, not laying in food for the winter,' her father taught her.

According to the first Austrian census of 1785, there were 526 Jewish families in the region. The numbers grew. According to the *Jewish Encyclopedia*, they worked mostly as 'craftsmen and owners of industrial workshops, tavern owners, moneylenders, builders, and real estate owners'. The better-off Jews spoke German and associated with the Germans of the town. The Austrian rulers from 1775 to 1918 even regarded Jews as Germans. According to the census of 1910, the 'Germans' of Bukovina made up over 21 per cent of the population – the third-largest nationality after Ruthenian (38 per cent) and Romanian (34 per cent). In 1904, 42 per cent of the students enrolled at the German university in Czernowitz (Chernivtsi) were Jewish. The poorer Jews were more religious, associated with the mystical Hasidic rabbis and were more likely to speak Yiddish.

When the Russian army invaded Bukovina in the First World War, many Jews fled, fearing persecution. The collapse of the Russian army in the Russian revolution of 1917 allowed Romanian troops to reoccupy the province in November 1918. All of Bukovina was incorporated into Greater Romania in the Paris peace treaties, along with Transylvania.

When the Molotov–Ribbentrop Pact was signed between Germany and the Soviet Union in August 1939, there were around 800,000 Jews in Romania, making up 4 per cent of the population. In June 1940, Soviet troops occupied northern Bukovina. The Romanian forces in the north withdrew without a fight. The Soviets also seized Bessarabia, including Moldova, going far beyond the secret protocols of the pact.

Romania joined the Axis powers, and in Operation Barbarossa in July 1941 Romanian and German troops reoccupied northern Bukovina. The Jews of Bukovina, despite their German affinities, were accused of having pro-Russian or pro-communist sympathies. Many were massacred in their homes and in pogroms on the streets, and tens of thousands were deported in terrible conditions to Transnistria. Up to 60,000 Jews were killed in Bessarabia and Bukovina by Romanian and German troops in 1941. And up to 120,000 Romanian Jews died in Transnistria. Paul Celan, a Bukovinian poet from Chernivtsi, took shelter at a friend's house, after pleading in vain with his parents to join him there. Most round-ups took place at the weekend. When he went back in the morning, the house was empty. His mother and father died in the labour camps of Transnistria.

Celan wrote in the beautiful German his mother insisted he speak at home, and was strongly influenced by Rilke and Hölderlin. Poems about oppression in the language of the oppressor.

> It's falling, mother, snow in the Ukraine:
> The Savior's crown a thousand grains of grief.
> Here all my tears reach out to you in vain.
> One proud mute glance is all of my relief . . .

Celan spent the war in one forced labour camp after another. He hauled the rubble of a destroyed bridge out of the Prut river and burnt Russian books at gunpoint, on the orders of his Romanian guards.

Sometime in July 1945 he hitched a ride on a military truck to Iaşi, then took a train to Bucharest and another to Paris. He never returned to Bukovina.

Some Jews survived the war. As a child in Gura Humorului in the 1960s, Lucy had many Jewish friends. But under another deal, this time between the governments of Israel and communist Romania, hundreds of thousands of Jews were 'sold' to Israel. 'In the centre of Gura there was a big Jewish community. There was a road called "Jewish street" up to the revolution. At one time there were five synagogues in Gura Humorului.'

She describes a visit to Gura by a group from Israel, led by a man called Oren Scharfstein. He managed to find eighteen relatives, from South and North America and Europe, and convince them to make the pilgrimage back to Gura. Lucy hosted them all, and they left a history of the Jewish community of Gura, with photographs and genealogies. 'The next day, one of the girls who helps in the kitchen found me sitting hunched over it, tears streaming down my face,' Lucy relates. They were all there, the girls she grew up with. The last Jews left Gura in 1972.

◎ ◎ ◎

Each of the monasteries of Bukovina has a particular colour. Each reminds me of a ship, a seagoing galleon, an impression strengthened by the broad brim of the roof built to protect the exterior paintings. The effect on the eye is to make the body of the church appear slimmer, like a boat, stranded in green fields.

Despite their size, most of the painted monasteries of Bukovina are in villages, not cities. They need a lot of space around them so that they can be appreciated in all their glory.

The monasteries were built after the fall of Constantinople to the Ottomans in 1453. The Ottoman advance transformed the architecture of church buildings: they now needed to be defensive, with fewer and smaller windows. Whenever Stephen the Great of Moldavia (1457–1504) won a battle, he ordered another church to be built to thank God for his victory. He fought the Hungarians and the Poles, too. And as he usually won, he had a lot of building to do. The builders, architects and icon painters who had fled Constantinople came looking for a job, and the magnificent monasteries of Bukovina were the result.

The Suceviţa monastery is enclosed by a square of stout defensive walls, 6 metres high and 3 metres thick. The Suceviţa river flows close by, and hills rise gently on either side. The vivid green of the monastery walls, decorated with Biblical scenes, strikes you the moment you enter through the big outside gates. Green is the colour of Islam, of Khidr – the green man, the mystic. This is my first green church.

The exterior paintings at Suceviţa and the seven other painted monasteries of Bukovina cover the whole wall, from top to bottom, like astonishing cartoon strips, depicting stories from scripture.

On the northern facade at Suceviţa, most endangered by the wind and snow, stands the Ladder of Divine Ascent, painted right up the wall – looking for all the world as though it had been propped up by casual workmen. This is a theme from the monasteries of Mount Athos in Greece, the well-spring of the Orthodox Church.

The story was written by John Climacus in around the year 600, in the Raithu desert near Mount Sinai in Egypt. It was addressed to monks, to teach them the straight and narrow path to Heaven – and to warn them of all the pitfalls that might cause them to fall. In the Middle Ages, the ladder was a popular image for illiterate Christians, depicting the struggles of life on earth and the hope of eternal salvation. It's a dynamic picture – long-bearded monks in long habits ascending the ladder bare-foot; hosts of angels flying overhead, urging them on; horned devils swarming below, trying to pull them off. The angels are like dragonflies, their haloes glinting gold in the afternoon light, their wings and tunics different shades of vermilion and scarlet. The space below the ladder is thick with falling monks. God welcomes survivors at the top of the ladder. Just below him, almost on the last rung, one white-beard has managed to stop his fall by hooking his feet around the top of the ladder. An angel is about to reach him. God is busy with the next saint. There are thirty rungs on the ladder, representing the thirty years of the life of Christ, each offering a teaching, a chapter of the original book. The first step is renunciation of the world. Number three is on exile and pilgrimage, and concerning the dreams beginners have. Number fourteen is my favourite: 'on that clamorous mistress, the stomach'.

In the 1960s, specialists carried out the first detailed analyses of the paints and techniques used, and the preparation of the plaster before the paints were applied. The first layer of plaster was a mixture of lime, straw and chaff. The second was made up of lime again, but this time with hemp fibre, used in Europe since Roman times, mixed with mortar in construction. Traces of *țuică*, the Romanian *pálinka*, were found in the mortar – too frequently to suggest the odd spill by drunken artists. It must have been an integral part of the recipe.

To apply this on the original walls of stone and brick, huge scaffolding was built, known as *ponte* – from the Italian word for 'bridge'. Analysis of the paintings shows that they were executed in horizontal bands, corresponding to the wooden planks of the scaffolding.

Simple geometric designs were added later, to cover the area where the plank rested against the wall, and to divide the rows of pictures in this 'comic-book' story.

Before the individual scenes were painted onto the lime and hemp base layer, the shape of the characters, their facial and clothing features, their beards and wings and toes, were incised onto the plaster. In Byzantine iconography, exact reproduction of the characters and the scenes was crucial, and there was little room for artistic licence. The facial expressions of the monks climbing the ladder are almost identical, as are those of the saints and the devils below. Only God himself, at the top of the ladder, preparing to welcome them into Heaven, is unique. His face is young, almost boyish.

This style of painting was developed in Constantinople on the great churches after the Latin sacking of the city in the Fourth Crusade in 1204. The Catholics murdered around 2,000 fellow Christians, destroyed irreplaceable items from the Greek and Roman period, looted their churches, and carried home booty, including the bronze horses that now stand in St Mark's in Venice. But the Venetian artists who lived in Constantinople during the six decades of Latin rule taught the art of perspective: buildings placed in front of other buildings; the different size of characters depending not just on their position in the Christian hierarchy, but also on their distance from the viewer. The monasteries of Bukovina were painted between 1488 and 1601.

Lumps of azurite for blue, malachite for green, cinnabar for red, and red and white lead came from all over Christendom and beyond. Much of the material must have been brought ashore at the ports of Chilia and Akkerman on the Black Sea coast. The word 'turquoise' originally meant 'from Turkey'. Charcoal black, black earth, brown and green earth and gold from the gold mines of Transylvania for the haloes could be obtained without long sea journeys.

The pigments were ground to dust by apprentices, then mixed and applied by the artists. Analysis of the interior and exterior murals at Voroneț, 'the blue monastery', show a high proportion of brick dust and sand in the exterior paints to make them more resilient.

Inside the Sucevița church, there is a prominent sign with a camera and a red line through it, which I surreptitiously ignore. But barely have I raised my phone for a second picture than a voice rings out. Sister Maria is initially fierce, and I imagine myself hurled from the ladder outside. But when she discovers that I am English, she relents. English being one of her many languages, an inspiring conversation follows, though she forbids me to record, note down or quote from it – strictly from humility, of course. All I can say is that her name is Maria and she has been at the monastery for fifteen years. We part as friends, the hint of a smile almost escaping from her black headgear. Abandoned as I am in the otherwise empty church, I even suspect that she leaves in order to let me photograph in peace. But I choose to believe instead that she trusts me not to do so.

Instead I take notes.

A barefoot God walks in a forest of green and red trees, his right hand silhouetted against a white rock, separating day from night. In golden robes and halo, he seems about to pluck a star from the sky. To the right of the altar are four fishes facing him, as he stretches his hand out towards them. It's the Fifth Day of Creation, when the waters were separated from the land. Over the doorway is an ox – the symbol both of Bukovina and of the wisdom of Solomon.

◎ ◎ ◎

In March 2022, I find myself back in Bukovina, just after the Russian invasion of Ukraine. I fly with colleagues to Bucharest, then drive to the Danube delta to watch refugees crossing the Danube by ferry at Isaccea. Then I head north, following the border with Moldova, through Suceava to the Siret crossing, the main escape route for those fleeing through Chernivtsi.

I contact the Orthodox Church to ask if they are looking after refugees. The press spokesman in Suceava is suspicious at first. The Orthodox Church in Romania has a difficult relationship with the media, and it takes a while to reassure him that my intentions are honourable. After a week, I am told we can visit the monastery of Sihăstria, near Putna, where two families of Ukrainian refugees have been given shelter.

Irina Babich, an English teacher from Kharkiv, is cleaning fish in the gloom of the monastery kitchen, with her friend Iryna. Her long, blonde hair is tied back, her shirtsleeves are rolled up to the elbows. There is a big crate of fish to go through, to feed many monks on the upcoming feast day. It's a medieval scene.

Later, we light candles with the women in a little chapel. A gleaming new church has been added beside the old timber one. The buildings fit into the landscape as if they have grown here. 'The other day, I found myself praying for Putin, and for the Russian soldiers,' Irina confesses.

She said goodbye to her husband, a fuel-tanker driver, packed her three small children and her mother-in-law – plus her friend Iryna and her two children – into the family's old Ford minibus, and drove the whole length of war-torn Ukraine to Romania. They had no idea where they were headed, except to safety.

Eventually, they found themselves at the Sihăstria monastery. Just up the road, Putna, the burial place of Stephen the Great, is much more famous, and draws the tourists and pilgrims. That leaves Sihăstria as a little haven of quiet and contemplation.

The snow is still deep on the ground, muffling the sound of the wooden board, beaten by a monk with a mallet, to summon the faithful to prayer. The women take us back to the clean, modern house that the church has given them for as long as they need it.

We hope and we pray here in the monastery that soon this crazy war will stop and this crazy man [Putin] will stop killing us and will stop destroying our native city. And we are sure that in the future Kharkiv will be better. And everybody in the world will come and see that we are happy.

Of course, we want to go back as soon as we can. I want to walk into my classroom. I want to say 'good morning . . . the topic of our lesson is Ukraine. The topic of our lesson is the present simple.' I want to dance with my pupils. I want to make concerts with them. And I want to go on picnics again, with all our friends in Kharkiv.

She shows us the inside of her 'beautiful van', with even the apple core – preserved in a plastic cup – that her husband left when he last sat there, a fortnight before. She cannot bring herself to throw it away. She bends down to sniff the steering wheel. 'I'm so glad that here I have a part of home, a part of my husband. Sometimes I want to sit in here just to smell my husband, to smell my house. The car is very old, but it's a very special car for us. I'm sure that soon I will go back home.'

We stay in contact. Easter comes and goes, then summer, autumn and winter. She doesn't reply to my messages, and I fear something terrible has happened.

In July 2023, I get a message. She and her family have returned to Kharkiv. They are all fine.

In the summer of 2024, as Russian advances again threaten Kharkiv, Irina, her husband and the children were given asylum in London.

CHAPTER 15

THE FORESTER'S TALE

The main crime of communism was the felling of the broad-leaved, deciduous forests of northern Romania and their replacement with monocultures of spruce.

Ion Barbu, Romanian forester

Ion Barbu – 'Bearded John' – is a big man. A forester with a passion for history. Born in 1953, he worked for more than forty years as a silviculturist – an expert on the science of planting, nurturing and cutting forests. White-haired, with a small, grey moustache and beetling eyebrows, he has a booming voice and speaks a colourful English, reinforced with frequent repetitions to underline certain words and phrases. When he finds a more accurate French or German word, that is thrown in for good measure.

I pick him up near his home in Câmpulung Moldovenesc for a tour of the forests of the Bukovinian Obcina, the northern slopes of the eastern Carpathians.

'The silver fir is the biggest tree in the forest of Europe. You can see trees 60 metres high, 2 metres in diameter and 40 cubic metres in volume.' I do a quick calculation. To heat my home in a west Hungarian village, I get through 3 cubic metres of wood each winter. So one ancient silver fir would keep my family warm for more than thirteen years.

The silver fir can live for up to 600 years, grows best at 500 to 1,500 metres above sea level, needs a lot of rain, and is sensitive to air pollution.

First stop, a clump of spruce on a high hill, the Obcina Mestecăniş ridge, above the sources of the Moldova and Suceava rivers.

The Suceava flows north from here, crosses into Ukraine, then runs along the border for a while, before rushing back into Romania. The Moldova flows south, giving its name to a whole country. Both are tributaries of the Siret, a river whose name in Hungarian means 'love'.

We walk across a pasture to the wood, fending off the attentions of some over-friendly cows. The trees are tall and slim, rather close together, all seventy years old and clearly sick. Some have collapsed, others have half fallen, propped up precariously on their fellows, like soldiers on the way home from battle.

Ion points to the badly scarred bark of a tree. More bark has formed in a protective ring around the wound, but too late. The tree has rotted through. 'This was caused by red deer.'

After the Second World War, forestry in Romania was taken over by Sovrom-Lemn, a Soviet–Romanian company tasked with ensuring that Romania paid its reparations to the Soviet Union. In post-war Romania, one of the few remaining resources was timber. Reparations were set at $300 million by Moscow. Although the countries ended the war on the same side, Romania's 1941–44 spell as an enemy was not forgiven. Whole forests, like the one which covered this ridge, were felled and the timber sent to the Soviet Union by train.

Sovrom engineers then faced the problem of what to plant in their place. The communists looked down on beech and oak as slow-growing and ungainly, compared to the speedy, straight wood of the spruce. Vast areas of beech and oak trees were wiped out. The wood was used in construction, burnt, exported as furniture, or turned into charcoal.

Fashionable as it is to blame the communists for everything, Ion says, they were not the first folk in history to favour the fast-growing spruce over the sturdy oak and fairytale beech:

> When the Germans discovered brown coal, they didn't need their broadleaved forests. They needed a resinous forest for construction in the mines, for pulp and paper. So they planted a huge area of former broadleaf forest with spruce and pine, which radically changed the characteristics of the soil.

The needles that fall under the pine are much more acidic than the decomposing litter of deciduous trees.

In the early 1950s, strychnine was used here to poison wolves and bears. The communist leaders of Romania loved to hunt, and venison was their favourite dish. In the early 1950s, red and roe deer were protected and their natural predators wiped out. The number of bears fell to its lowest level in 1953 – just 860 in the whole country. The resulting surge in the deer population was a disaster for the trees that the authorities had begun to plant by the hundreds of thousands, to replace those cut to satisfy the hunger of the Soviets.

Spruce were sown by scattering cones on the snow in the late winter. The cold broke them apart, and as the snow melted in the spring, the seeds fell neatly into the undergrowth, the fertile soil left by the deciduous forests that used to stand here. The saplings grew fast, basking in the summer warmth. Then the deer arrived and began chewing their bark. The trees, to heal the wounds, grew more wood around the scars, but the damage was done. In the long, damp winters, fungi penetrated the wounds and gradually rotted the trees from within. Now seventy years old, this forest is at the prime age for harvesting. Instead, the wood is good for nothing. Strychnine was banned in 1968.

Another problem is the sheer density of the trees. They should have been thinned out long ago:

Our forests are too dense, and too dense is risky. A forest is like a glass. I pour in water or wine, and there comes a point when it is full. One of those limits is the competition between the crowns of the trees. Every crown wants to be in the light.

While the environmentalists say Romania is cutting down too many trees, Ion argues that the country is planting too many, too close together.

We drive down to the source of the Suceava river at Izvoarele Sucevei. When the Germans arrived here in the late eighteenth century, their first complaint was about the food: the Romanian shepherds ate mostly mutton, but the Germans and Austrians needed beef and veal.

The *Wiener Schnitzel*, a slice of calf meat rolled in flour, then fried in egg and breadcrumbs, was first mentioned in an Austrian cookbook in 1831; but as *Kalbsschnitzel* it had long been a sought-after delicacy. Herds of cows were driven across the mountains from Galicia by the Hutsuls to satisfy the hunger of the newcomers.

With so many languages bandied about, the Romanian spoken in Bukovina is somewhat different from the vernacular in the rest of the country, with many loanwords from the languages of the settlers. One is the German *zurück* – meaning 'get back!', as shouted at horses, though now used mostly to refer to the reverse gear on a car. The purple hills of Ukraine jostle for position along the horizon. This meadow is the source of four rivers – the Suceava, Moldova, Aluniş and Sludnica.

A vehicle screeches to a halt beside us. 'Hey, professor!' A former colleague accosts Ion. The man retired recently and both his sons studied with Ion at the forestry school in Suceava. Another man, so drunk he can barely stand, staggers up, angling for a lift down the valley. But we're going up it, towards the Ukraine border. We park the car beside the Romanian border police post. It's hard to believe that this will one day be a border crossing between Bukovina and Galicia, with trucks and stray dogs and chain-smoking customs officers. They've been talking about opening the border here for twenty-four years, says Ion.

In the tiny hamlet of Pârâu Negrei, near Breaza, we visit a collection of traditional Hutsul timber houses, dismantled and reassembled as a living museum for tourists. The long terraces provide shade from the generous Bukovinian sun. Bright red geraniums hang from the eaves, and the window boxes are full of carnations, the green leaves and bright flowers contrasting with the dark wood. The roofs are steep and shingled.

Inside, the rooms are low-ceilinged and lavishly decorated with embroidered linens. In the kitchen, a white stove takes up most of the space, with a range of warm and hot ovens. There's a black-and-white photograph of the original inhabitants, a stocky peasant woman in an embroidered smock and her groom. Electric light sockets are concealed in the timber, and the horse-hair mattresses have been replaced with something more modern. There are even bathrooms – unimaginable luxury for the original inhabitants.

Negrei (in the name of the hamlet) refers to the black local sandstone. The timbers of the houses are built on it, to lift them off the ground in the long, wet winters.

We cross the ridge to Lucina, at 1,500 metres the highest peak in this massif, home to a famous stud of Hutsul ponies that dates back to 1792, soon after the arrival of the Austrians.

The ancestors of the Hutsul horse appeared in these parts with the Tatar invasion of the fourteenth century. The tarpan, *Equus ferus ferus*, a wild Eurasian horse, is now extinct. The last known animal died in captivity in Russia in 1909. Hutsul ponies are most likely the result of interbreeding between the original tarpans and local horses. Hutsul ponies were discovered by the Austrians as an ideal horse for the tough Carpathians, and five studs were opened, including the one at Lucina.

It's a single-track, rutted road, and we swerve often to allow trucks or madmen in BMWs to pass. The road at Lucina is an alleyway of silver firs, each painted white at the base. Above the track, stallions wait impatiently for the mares to be brought to them. It's mating season, and Costica, the man who shows us around, says the males would fight each other to the death if they were not kept strictly apart. There's a palpable tension in the air as we enter. The horses watch us, wild-eyed through their long, black manes. Accustomed to smaller Hutsul ponies in Ukraine and Slovakia, I'm surprised how large these seem, stomping and snorting around their green wooden stalls. Each horse has a name and number, identifying it in the bloodlines established here since 1856: Goral, Hroby, Ouşor, Prislop and Pietrousu. The first we visit is Ouşor 14, a mousey-coloured horse, next to Ouşor 13, black as night. Then Prislop 12.

Costica is from Benia. Hutsul horses can be dark bay or chestnut, mouse-coloured or black. They have large, almost oversize-looking heads, short, tough legs, and stout barrels for bodies.

The stallions can pull carts loaded with up to 2 tonnes. That has made them useful in warfare, as well as farming. The stock was devastated in the Second World War on the battle fronts of Transylvania, when they were used to pull artillery pieces and munitions. Many ponies are also bred here for sale abroad. The Hutsul ponies – except when mating – are generally amiable and friendly to humans, endearing them to children. Their broad backs and sturdy footing also make them ideal for horseback archers.

The first horse remains found in the Carpathian basin date to around 3,500 BC – the Copper Age, around the same time as horses were first tamed on the steppes of Asia. There is an image of a horse drawn in black charcoal among half a dozen pictures of animals found at the Coliboaia cave in the Apuseni mountains, dated to between 23,000 and 35,000 BC.

Mr Lesenchuk, the blacksmith who shoes the horses and prepares them for competitions, has worked here for forty-one years and is on the brink of retirement. But he will keep doing the same work, he says. 'I have a smithy at home, just like this one. A horse, a few goats, and my own company – I'm independent!'

And who will take over his job at Lucina? He shrugs. 'Everyone is just interested in getting a diploma nowadays, but you can't learn a trade like this at school.'

Three hundred mares and fillies are driven down from the hills at dusk, to their night quarters in the valley.

Ion has kept the crowning glory of our day till last: the primeval forest in Slătioara, more than an hour's drive from Lucina. He defines 'secondary forests' as areas where trees with a diameter of between 28 and 80cm were cut, prior to the First World War. Younger and older trees were left, and locals made use of the ancient, partly rotten trunks to make shingles to cover their roofs. Only the central part of the tree rotted away, leaving a solid, hollow trunk. That wood split very well.

Romanian forestry has been changed in the past decades by the arrival of three large Austrian firms: Schweighofer, Egger and Kronospan. The forestry industry, starved of capital after the revolution of 1989, received a massive shot in the arm. Processing of timber shifted into high gear.

Ion admits that he was upset by the impact of the Austrians, but says there was no alternative:

> They invested strategically, because they speculate on the price of wood between Ukraine, Belarus and Romania, so they always buy for the lowest price. It is good that Kronospan and the others are here, because they use our raw material. Without them there would be no market. But they use very few workers. Engineers to select the best wood to buy, at a very low price, and they get all the profits for themselves. They leave nothing in Romania. When we encouraged them to come here, they promised workplaces, high salaries and sustainable development for the local population. They have not kept those promises.

He laments the disappearance of the Romanian furniture industry in the 1990s. His friend was the director of the big furniture factory in Câmpulung, which employed thousands. Their oak furniture was famous, especially on the German and US markets. A shopping mall has been built on the site of the former factory.

Ion takes issue with the environmentalists who campaign to save the forests of the Carpathians:

> They always say: don't touch the trees! Leave the deadwood in the forests! But this is not realistic. For the past 200 years, all the forests were organised to produce wood. Now people come along, Romanians and the European Union, and tell us, no, we have to change the purpose of the forest. But it would be very risky now, to stop all the cutting.

His vision is summed up in the German word *Durchforstung* – 'thinning'. This would allow companies and local people to regain access to the forested mountainsides of their homeland, from which they have been excluded.

> If there are 5,000 trees per hectare at 10 years, we must have a maximum of only 500 at 100 years. During those 90 years, I must extract 4,500 trees! If I do not, nature will eliminate them by herself – because nature has her own strategy.
> We need a wood industry, but not like Schweighofer or Kronospan or Egger. A very diversified industry which uses all the qualities of the wood. Pulp and paper production is very important because it uses materials of low quality. We need a domestic furniture industry again, and a domestic cellulose industry.

In communist times, there were three big factories in Romania producing cellulose – a kind of industrial, man-made silk. One was in the regional capital, Suceava. All three have closed down.

In recent years, the Austrian companies have tried hard to improve their public image, in response to critical media reports: 'We are proud to work with a raw material that offers a way forward in combating the climate crisis,' writes H.S. Timber on its webpage.

On the way from the Obcina Feredeului to Slătioara Ion and I discuss the management and mismanagement of Romanian forestry.

One major problem, he suggests, is that ministers of the environment are selected according to political rather than expert criteria. I mention the names of Barna Tánczos, from the ethnic Hungarian party, and his predecessor Costel Alexe. 'Both were baked from the same flour,' he says picturesquely.

And what about President Klaus Iohannis? He said recently that the sheer extent of illegal logging in Romania was 'intolerable'.

'What is tolerable and what is intolerable?' Ion asks rhetorically. As existing trees grow, 'there is a natural increase in the wood volume of Romania of between 50 and 60 million cubic metres a year. If the forest is well managed, we could afford to cut 50 million cubic metres a year!'

I describe what I had seen in Valea Ursului, where Romsilva had cut almost all the trees. He knows the story and denies nothing. He offered to draw up a map of all the clearcuts in Romania from 1990 to 2000, he says, but there was no interest from the Romanian Academy. In the preparatory fieldwork, using Google Maps alone, Ion and his students were able to establish that there had been far more damage caused to the forests on the Romanian side of the border than on the Ukrainian. The map was never published.

Satellite images published by the Humboldt University in Germany are frequently cited to prove the extent of the destruction of forests. Ion says that such images should be followed up with field trips to study the situation on the ground. Many such areas, he suggests, are the result of natural disasters like windthrows and bark-beetle attacks, and not the consequence of over-zealous or illegal cutting.

There have been several attempts to record and map all the primeval and quasi-primeval forests of Romania. The 'Pin-Matra' study conducted by Ion's National Forestry Institute from 2000 to 2004 estimated between 217,000 and 250,000 hectares. In 2017, Greenpeace Romania suggested closer to 300,000 hectares, because important forests had been left out of the original survey. Greenpeace, the WorldWide Fund for Nature (WWF) and Agent Green all claim that there is indecent haste on the part of foresters to cut as much as possible, because they know that, sooner or later, they will be stopped.

According to Ion:

> The buffer zone is not an untouched zone. It has been managed in the past hundred years by men using different methods. So if we forbid any interventions there, we will have a lot of dead trees which remain in the forest.

He agrees with the ecologists that the optimum forest is one of trees of different species and ages. 'Wind is the most damaging factor. In even-aged forests, exposed to extremes of wind, all the trees are damaged at the same time.'

We arrive at Slătioara. Some explanatory panels, a couple of low buildings where visiting forestry scientists can stay, but otherwise little sign that we are entering one of the most treasured places in Europe.

We set out on foot. A mixture of beech and fir towers overhead as we follow the path. The mood in the forest changes, from the everyday bustle of trees and birds to something deeper, as though the trees are holding their breath.

'This is very interesting,' Ion says, pausing at a tall sycamore by the path.

> The new trees growing here are more dominated by the broadleaves. This grows wherever it finds enough light, when a gap in the canopy appears. This means that the climate has become more and more favourable for this sensitive species.
>
> This is nature. Nature expends a lot of energy, which is first accumulated in wood, and then transformed into deadwood, so the other trees can thrive. But the strategy of men is totally different. Why let such wood decay, when I can use it, give it a price, and a value?

We enter the virgin forest through a sort of natural gateway formed by two silver firs. Our footsteps echo. Each tree seems taller than the last. They are mostly well spaced and uncrowded, except where one has fallen. The only sign of human intervention is the path we follow, deeper and deeper into the forest. Where a tree has fallen across the path, a ranger with a chainsaw has cut a way through.

We study a fir, 50 metres high. Since the 1980s, he and his colleagues have established sample plots here, describing each tree, its age and condition, in a radius of 500 metres. Every five years, the study is repeated. 'This is what is known in German literature as the *Zerfallsphase*' – a period of decay, when large trees fall, but remain important on the forest floor, sustaining the life of a myriad fungi and smaller forest insects. Again, the Germans have an appropriate expression: *Kadaver-Verjüngung* – 'regeneration of the corpse'. While most of the older trees in the forest are evergreens, the younger ones coming up are broadleaved. Then we reach a silver fir he estimates at 500 years old.

I take Ion's picture by the tree. We are both grinning with childlike pleasure. South of here, in the Călimani mountains, there are stone pines like those I saw in the Kôprová valley with Martin Mikoláš.

The dwarf pine grows above the ordinary treeline, up to around 1,900 metres. 'The communists had a plan to destroy that, too, by burning, in order to create more highland pastures.' Some of the plan was put into effect, and Ion has doctoral students studying the recovery of the land from what he calls 'that barbaric practice'.

'A lot of mistakes and arrogant methods were used in the past. But we must learn from past mistakes, or risk repeating them. And society is not open to support such costs.' German, Polish and Czech ecologists have been here in search of insects that haven't been seen in their own forests for more than a hundred years.

We come to a spring named after Franz Cech, director of forestry in Bukovina in the 1930s. Thanks to him, Ion says, Slătioara still exists today. He went to Bucharest to petition King Carol II against plans to cut it down. The spring water tastes delicious.

We take off our shoes and socks to walk slowly back through the forest. Early in his career, Ion was tasked by the Ministry of Health with producing a scientific report on the health benefits of forests:

All plants produce negative ions, while computers, mobile phones, much of what we surround ourselves with in our cities, produce positive ions. A forest like this with a canopy more than forty meters high produces a huge amount of negative ions. We suffer from stress, all the time we are isolated from the earth, and we become collecting points for positive ions. By walking barefoot in such a place we expose ourselves to the negative ions which neutralize the others. That's why we feel so wonderful when we walk through this forest, or in the high mountains, or beside the ocean.

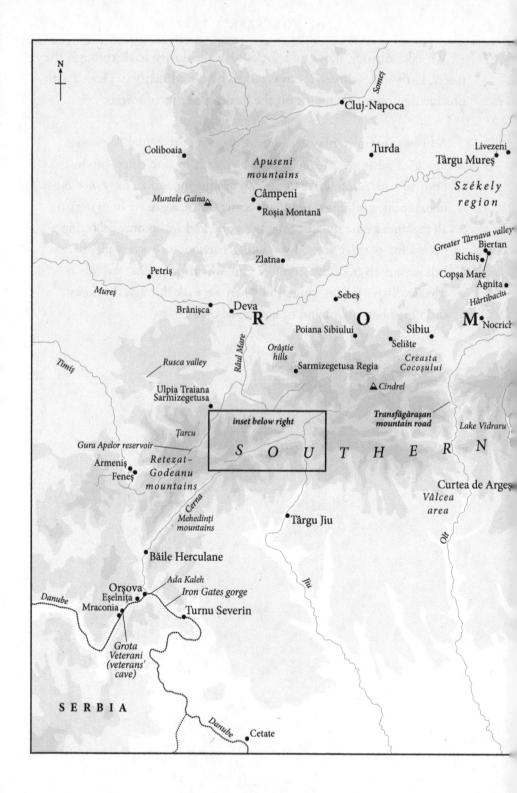

N

Somes

Cluj-Napoca

Coliboaia

*Apuseni
mountains*

Turda

Livezeni

Târgu Mureş

Muntele Gaina

Câmpeni

Roşia Montană

*Székely
region*

Greater Târnava valley

Biertan

Richiş

Zlatna

Copşa Mare

Agnita

Petriş

Hârtibaciu

Mureş

Sebeş

Brănişca

Deva

R   **O**

Poiana Sibiului

Sibiu

M

Nocrich

Selişte

Timiş

Rusca valley

*Orăştie
hills*

Sarmizegetusa Regia

*Creasta
Cocoşului*

Ulpia Traiana
Sarmizegetusa

Răul Mare

△ *Cindrel*

Ţarcu

inset below right

*Transfăgărăşan
mountain road*

Lake Vidraru

Gura Apelor reservoir

S   O   U   T   H   E   R   N

Armeniş

*Retezat–
Godeanu
mountains*

Feneş

Curtea de Argeş

*Vâlcea
area*

Cerna

*Mehedinţi
mountains*

Târgu Jiu

Olt

Băile Herculane

Ada Kaleh

Orşova

Iron Gates gorge

Jiu

Eşelniţa

Mraconia

Turnu Severin

Danube

*Grota
Veterani
(veterans'
cave)*

S E R B I A

Danube

Cetate

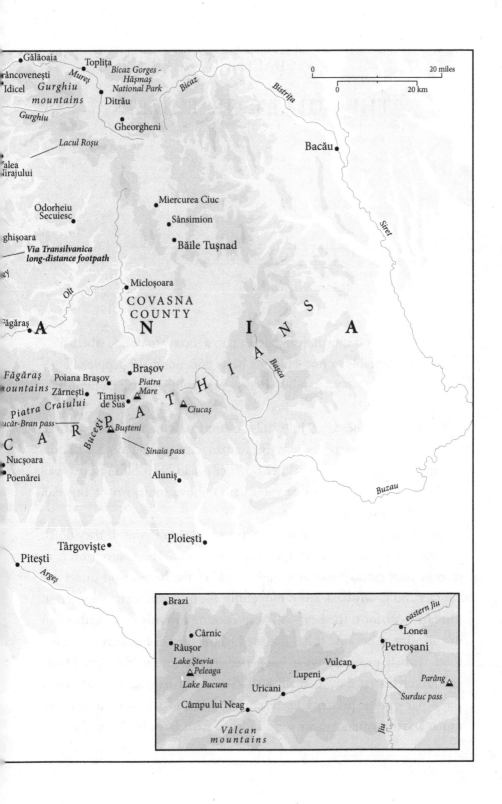

Gălăoaia

Toplița

Brâncovenești

Idicel

Mureș

Gurghiu mountains

Bicaz Gorges - Hășmaș National Park

Bicaz

Bistrița

Ditrău

Gurghiu

Gheorgheni

Bacău

Lacul Roșu

Valea Mirajului

Siret

Odorheiu Secuiesc

Miercurea Ciuc

Sânsimion

Sighișoara

Băile Tușnad

Via Transilvanica long-distance footpath

Micloșoara

COVASNA COUNTY

Olt

Făgăraș

A

N

I

N

S

A

Brașov

Bâsca

Făgăraș mountains

Poiana Brașov

Zărnești

Piatra Mare

Piatra Craiului

Timișu de Sus

△ Ciucaș

Bucăr-Bran pass

R

P

A

T

H

Bușteni

C

A

Bucegi

Nucșoara

Sinaia pass

Poenărei

Aluniș

Buzau

Ploiești

Târgoviște

Pitești

Argeș

Brazi

Cârnic

Râușor

eastern Jiu

Lonea

Petroșani

Lake Ștevia

△ *Peleaga*

Vulcan

Lake Bucura

Lupeni

Uricani

Parâng △

Câmpu lui Neag

Surduc pass

Jiu

Vâlcan mountains

CHAPTER 16

THE BULLET OR THE BIN

All hunting is trophy hunting. It is up to a society to decide whether
that is acceptable or not. Hunting bears does not make people safe.

Julius Strauss, bear guide, British Columbia

At 15.00 on Tuesday, 9 July 2024, Diana Cazacu, a nineteen-year-old
student from the eastern Romanian city of Iași, rang 112, the
emergency number. She and her boyfriend had been hiking up a trail
on Jepii Mici ('small juniper mountain') above the town of Bușteni,
when they encountered a bear.

Bears, on the whole, are afraid of people and run away. If they don't,
you have two choices, both (potentially) correct. The first option is to
stop in your tracks, turn at a slight angle to the bear, avoid direct eye
contact and slowly back away. Hopefully, the bear will then lose interest.
On rare occasions, the bear will keep coming towards you – either out
of curiosity or because it has got used to the idea of humans providing
food. This is probably what happened in the case of the bear Diana
encountered that day, a smallish, 4–5-year-old female. In that event,
the other option presents itself: confront it and try to scare it away,
much as you would a stray dog. In areas with large bear populations,

like the Bucegi mountains, you should carry pepper spray for precisely this eventuality.

The one thing you should never do, bear experts agree, is to run away. Because bears, like dogs, will instinctively run after you, even if they were not particularly interested in you when you first spotted each other.

Tragically, Diana panicked and started running back down the path. The bear ran after her, knocked her to the ground and bit her leg. Diana's last words, recorded by the emergency services, were 'It's attacking me! It's attacking me.'

She fell down a ravine, and the bear climbed down after her. When the Salvamont mountain rescue team reached the spot, they found Diana's body badly mauled, her head in the stream, and the bear standing over her, apparently eating her. The bear then attacked the rescuers, who fought back with pepper spray and fire crackers. Then Doru Busijoc from the local hunting association arrived and shot the bear.

Sergiu Frusinoiu was the team leader of the Bușteni branch of the Salvamont on duty that day. I take the cable car in Sinaia, up to the Salvamont headquarters at Cota – a modern, two-storey building bustling with men in red-and-white uniforms, with maps and climbing equipment. Nine people died in this region in 2023 alone, he says – mostly falling from steep cliff faces, or exposed to the elements when they didn't bring the right gear. Most of the injuries are in the summer months, when the mountains are most crowded – the Bucegi are only a few hours on the main road from Bucharest.

Diana, he says, was ill-prepared for such a hike, wearing city shoes. It's a moderately difficult hike, which tourists enjoy as a challenge, for the beautiful waterfall and the fine views from the plateau above. Her death has sent shockwaves through the rescue service across Romania. There's an internal debate in the Salvamont about whether their protocols for dealing with bears need to change, as the bears change.

Based on the experience of hundreds of years, we know that bears don't eat humans, they don't consume human flesh. They do kill humans, however, if they think they're in danger.

Sergiu is thirty-two and has a tattoo of a magnificent silver fir tree on his right forearm. His father came originally from the plains of southern Romania and was fond of hiking. 'The mountain people, those born and bred here, don't do much walking in the mountains. But my father, as an outsider, was keen to explore.' Sergiu's earliest memories are of long hikes with him, from the age of five:

> He taught me that so long as the bear leaves me alone, I should leave him alone. He also taught me to live in accord with nature, not to try to be 'more' than nature. These mountains have a special energy. And because they're so steep, they're very wild.

He started as a tourist guide, but switched to rescue work six years ago. He wouldn't trade his job for anything in the world, he says, and gets homesick after more than three days away.

He reckons he has had more than a hundred encounters with bears, about a third of them 'close encounters', less than 5 metres away from the bear. 'They were never aggressive, even the mothers with cubs. What happened in this latest case is very strange and unusual for me.'

From Sinaia, I drive to Bușteni, to the site of the accident. At the base of the footpath, looking up towards the Jepii Mici peak, the roots of a tall fir tree offer footholds to hikers, and there's a prominent 'Attention, Bears!' sign nailed to a tree. The trail Diana and her boyfriend took is well marked, with blue crosses on a white background. At the base of the footpath is a large car park, behind the Hotel Silva, and three large municipal waste bins, surrounded by a cage to keep away wild animals. But the cage is wide open, and a large yellow bin for food waste has been tipped over, presumably by bears. Food waste, plastic packaging and tins are strewn over a wide area, furtively grazed on by stray dogs. Clearly, the bear that attacked Diana was on its way here. The failure of the municipal authorities in Bușteni, and in so many places in the Carpathians, to tackle their own waste issues, is at the root of Romania's 'bear problem'.

That evening, I'm cruising round the town, looking out for bears, when an 'Extreme Bear' alert screams from my phone, and a text message flashes up, in Romanian and English:

> The presence of a bear was reported in Bușteni, Arcului street! Avoid the area, stay indoors! Stay away from the animal and do not try to take pictures of it or feed it. Protect your pets/livestock without putting your life in danger.

I drive immediately to Arcului street, a quiet residential road with family houses on either side, on the upper edge of the town, at the foot of the mountains. A small cluster of residents are out in the street, and I stop to talk to them. The bear, it seems, was spotted walking past a few minutes earlier. They're afraid to walk home after dark, they tell me, because of the bears. A police patrol car pulls up.

'Where have you been?' an elderly man shouts at the two hapless young officers. 'What exactly do you want us to do?' asks one of the policemen. 'Protect us from the bears!' shouts the man. I start to feel rather sorry for the police, but when I wonder aloud what in fact they can do, the angry man turns on me. 'I don't believe you're even a reporter!' he shouts, excitedly. But he offers no clue who I might be, instead.

Within a week of Diana Cazacu's death, the Chamber of Deputies in Bucharest passed a law allowing the hunting of nearly 500 bears by the end of 2025, doubling the previous ceiling. The new law also permits foreign trophy hunters to shoot bears in Romania. This means that hunters can go into the mountains actively looking for bears to shoot them in their own habitat, on the grounds that there are 'too many'. This is more or less a reversion to the pre-2016 situation, when hunting also failed to solve the problem. There is little political discussion of waste or recycling. It's as if the politicians think it's somehow 'beneath' them. The bullet is sexier than the bin.

Romania seems torn between a love-affair with bears and a hate-affair, with the tabloid media feeding both extremes. TikTok and other social media are full of video clips of cute bear cubs on their hind legs, begging at car windows on the Transfăgărașan highway, and biscuits and other food being thrown out to please them.

I meet Barna Tánczos, from the Democratic Alliance of Hungarians in Romania (UDMR), the main author of the law to restore hunting quotas, at a burger bar on the outskirts of Csíkszereda (Miercurea Ciuc).

> It's a huge achievement that after almost eight years [since the 2016 ban] we have a law which can control the bear population in Romania, which is increasing daily, monthly, yearly. If we don't do something we will have thousands, tens of thousands of bears, which is not good for humans, and not good for the bears. We need to reach a balance in human–bear relations and conflicts.

He doesn't hunt himself, he explains, though wild animals are hunted all over Europe, and that is normal. His critics have accused him of being a pawn in the hands of a hunting lobby which wants to restore bear-hunting at all costs. The lobby has also been accused of 'orchestrating' calls to the emergency services, to exaggerate the danger.

If the new law fails, will he try to increase the quota of bears that can be shot?

> I think this is the end, not the beginning of a process. Of course the law can be modified and adjusted. But now we have taken the toughest and biggest step. There is no risk for the bears. The bear population is in a good condition in Romania. We have the best brown bear population in Europe.

Cristi Papp, head of the large carnivore department of the World Wide Fund for Nature (WWF) in Romania, meets me at a filling station near the airport in Cluj. We've spoken many times over the years, in the forest or the city. He's a softly spoken man, with a passion for the wilderness. He monitors the bears, wolves, lynx and jackals that roam these hills. Romania is in the midst of an era of unprecedented urban expansion and motorway building, which is encroaching on the last strongholds of the large carnivores. The new motorways cut across bear migration routes, and the spread of towns and cities brings more people into contact with bears. Worst of all, a deeply ingrained custom in Romania of feeding wildlife has led bears to regard humans as providers of food.

There are indeed individual problem bears, who can be shot if they are proven to be dangerous. But they have become problematic because of past bad practice by the authorities.

We should aim at the root causes of the conflicts. And there are many causes, starting with feeding of the bears, which is happening massively close to the settlements, and it's happening along the roadsides in the touristic areas. Another problem is that waste management is not properly dealt with in many communities. We need intensive education campaigns and public awareness campaigns.

Before the 2016 hunting ban, local hunting associations were responsible for estimating the number of bears in Romania, and were allowed to kill 10 per cent each year – so an estimate of 6,000 bears would mean 600 could be hunted. Foreign hunters, especially Germans, paid the hunting associations up to €20,000 for the opportunity to kill a big alpha male. There was a strong financial incentive for the hunters to exaggerate bear numbers.

According to Cristi, the ban allowed a breathing space for wildlife groups like WWF to organise their own count of bear numbers – with tracking, scat samples, observation platforms and a small number of GPS bear-collars. One particular bear – a 3–4-year-old male known as 'Charlie' – crossed thirty-eight 'hunting districts' between Târgu Mureş and the Ukrainian border: during this time, he could easily have been 'counted' and reported a hundred times. Charlie crossed the border 'illegally' into Ukraine, and disappeared from the satellite, probably shot by poachers.

Bears can migrate hundreds of kilometres to find a mate. If they are trapped in a small pocket of land by motorways, there is a danger of inbreeding. An environmental impact assessment (EIA) should take place before any major investment, including the construction of roads. Wildlife bridges can be built over motorways. There's a particularly successful one at Mengusovce, between the Low and High Tatras in Slovakia, which wild animals use in droves. During construction of the Transylvania motorway, negotiations with the companies concerned led to three such bridges being built, on ecological corridors. Wherever they were not built, around the cities of Sibiu and Sebeş, the number of accidents involving bears, wild boar and deer grew dramatically.

'Unfortunately, the big interest groups which plan infrastructure projects pay for the EIAs themselves, and they only pay if they get a favourable one,' says Cristi. The WWF has developed an app to register roadkill; the organisation sends the data to the Ministry of Transport, emphasising the danger if no eco-bridge is built. It also provides footage from hidden cameras to show how much animal traffic there is – given the opportunity.

The debate on how to manage the bears is a complex one, not helped by the fact that successive governments have shown little willingness to sit down with biologists, conservationists, foresters and hunters, or to launch a public education campaign to discourage tourists from feeding the bears, and understand how to behave if they encounter a bear in the wild. The point is not how many bears there are, but how wild they have managed to remain.

With my friend Mircea Barbu, we arrange to spend a night in a bear hide run by János Szin, a tour operator dealing especially with bear tourism, near St Anna's lake, 6km from the village of Bixad. We're told to be at a roadside parking spot at exactly ten past six. We get there a little early, and several other cars arrive with Romanian and Dutch tourists. A man in a big green van with an URS – 'bear' in Romanian – registration plate arrives right on time. We're the only visitors who are planning to spend a whole night in the hide, which is raised above the ground, about five minutes' drive away down a mountain track. The others will stay in a separate hide at ground level, and watch the same bears.

Because of the big number of bears – around thirty regularly frequent this place – we're under strict instructions not to leave the hide until eight in the morning, when the ranger will come to pick us up. We reach it by a steep ladder, up onto a small balcony or viewing platform, partially hidden beneath a beech tree. Inside, there's a single large room, with a double bed, comfortable armchairs, rugs and a small wooden table. There's electricity from a solar panel, a tiled stove and a small bathroom. The owner normally charges €300 a night. He's let us stay for free, because he wants me to write about bears – 'but you can write what you like,' he says, when we make the arrangements beforehand.

At 18.30 sharp, the forest ranger drives up. Mircea and I watch through a large glass window in our room. We count five or six bears already loitering, waiting for feeding time. The ranger takes a sack out of his car, shouting at the bears to stay back. He doesn't seem in the least afraid, more like a secondary school teacher serving food to teenagers on a summer camp. He scatters most of the maize in a line, leading right up to our window, but places some in a black steel trough under a fir tree. One of the bears, a female, jumps up onto that immediately. The trough is connected to a digital scale, just above our window. She weighs 166kg.

For the next hour and a half, the bears munch happily away, occasionally growling at each other. At one point I crawl out onto our balcony to put a sound recorder on the ledge. Then a large male arrives. Perhaps he was here all along, and wanted his family to eat first. In any case, the mother with the three small cubs ushers them away. Another female bear keeps grazing on the maize with the big male. At one point, he climbs up onto the scale to feed there: 265kg. Soon after eight in the evening, just as its getting dark, the rain starts. The storm breaks over the mountains, with flashes of lightning and a heavy downpour that lasts most of the night. I imagine the bears, sheltering together, deep in the forest.

I wake at 05.00 to a clear dawn, followed by a watery sunrise. Jays are squabbling noisily over the last grains of maize left by the bears. A young female bear appears, and the birds and the bear work side by side, clearing the grass, as if in partnership. At six, when the bear has gone, Mircea flies his drone from the balcony – a delicate manoeuvre, with so many trees around. And we can finally see where we are, embedded among the dense forests of Transylvania, with distant villages just clearings in the woods, gossamer clouds still clinging to the low valleys, rounded mountains shoulder to shoulder.

On the phone, I speak to János, who runs the hide. Alone of all the bear experts I speak to, he maintains that there are fewer bears than before in Romania, not more. The public perception is fuelled only by 'bin-stories' he says, and has little to do with the bears up in the mountains. Male bears frequently kill cubs to get them out of the way to mate with the mother, rather like lions, he explains. The bear population of Romania has probably fallen as a result of the 2016 ban, he believes, because of this practice – there are many more big males in circulation.

◎ ◎ ◎

Băile Tuşnad (Tusnád) was propelled to glory by the construction of a railway station in the late nineteenth century, and the new popularity of 'taking the waters'. Beautiful old villas dot the town, their wooden eaves imitating the peaks of the mountains, their wooden terraces daydreaming of an era when people still had time to chatter idly and play cards in the shade on a summer's evening.

The railway line runs parallel to the River Olt, and the valley gives the plaintive blast of the train horns a special quality, the sound lingering longer than it should, bouncing between the wooded slopes overlooking the river. The town's other claim to fame is the annual gathering of Viktor Orbán, the Hungarian prime minister, and his supporters here, in the campsite next to the Olt. First held in 1991, as a meeting of Hungarian and Romanian intellectuals to heal the wounds of the ethnic clashes of March 1990, the Romanian component has slowly evaporated.

András Lakatos is waiting for me in his car outside the guesthouse where Mircea and I are staying in Tusnád. He's thirty years old, with a son aged three and a daughter of four months. He worked first as a forester, then a forest guard, patrolling 1,600 hectares. 'My everyday task is to care for the wild animals, the feeding, the observation of their habitats, and to reduce the poaching.' There are twice as many bears as there were when he grew up, he reckons. He has had numerous encounters with them, all benign. Once he came across the entrance to a cave in the undergrowth, by accident, just as the bear was coming out. 'It was fine,' he laughs. 'She ran off down the hill, and I ran up it!'

Another time, he had a close shave with a mother bear and her cubs. 'The bump on her back went up, a sign that she was preparing to attack,' he says. 'But I backed off, slowly and carefully, and she had second thoughts.'

Nevertheless, there is a bear problem in Transylvania, he believes, and he's convinced that the bears need to be 'managed' – a return to shooting a limited number.

'We didn't always feed them. It's just that humans have spread through the animals' habitat, we took away their living spaces, we built railways, developed agriculture, carved out fields where forests used to stand, and the settlements grew and grew.' The feeding of wild animals was introduced to try to keep them at a distance, away from the settlements. It's all carefully regulated, he explains: which food for which animals, how often, and where it is placed. Carrots and potatoes for the deer, alfalfa for the wild boar. They used to feed the bears animal carcasses in communist times, but that is banned now, as it gives them a taste for blood.

András has never heard of a single case of a bear going after a human in order to eat them. About 95 per cent of conflicts could be avoided, he reckons, if people were more sensible.

He's concerned about the spread of bear hides, however, like the one we stayed in. 'You could have one or two such viewing points per region,' he says. 'But we couldn't have hides everywhere. That would turn the bears into feral dogs, much worse than now.'

Bears have an excellent sense of smell, and get used to the forester who puts the food out for them. But when you have a tourist viewing point, the bear sniffs out a constant turnover of new people and gets confused. A strange smell used to arouse her suspicions and warn her of potential new danger. Now she's got used to humans providing her with food. The absence of shooting has bred a new 'complacency' in the bears, András reckons.

Zsolt Butyka is the mayor of Băile Tuşnad. With a bearlike hand-shake and a solid beard, he wouldn't look out of place on a big motor-bike. There's a Hungarian majority in the town, but he also speaks to the sizeable number of Romanians. That also means he doesn't just go to Budapest with his begging bowl, but to Bucharest as well.

For many years, Tusnád was famous in Romania for its bear problem: bears would wander the streets raiding bins and fruit trees, as they crossed the valley between the Bodoc and Harghita mountains. In 2021, there were 220 bear alerts; by 2023 this had fallen to just six.

'We brought in a team of experts from the WWF, and from the Babeş-Bolyai University in Cluj,' the mayor explains, 'to study the main problem districts in the town.'

In response to their findings, the mayor's office spent €10,000 on fourteen special bear-proof bins from Slovakia, cut down the apple and plum trees in the town centre that were attracting the bears, and replanted non-fruiting trees instead. A public education campaign also helped, he says.

'The problem is now 90 per cent solved.' The bears still cross the river here, 'but they avoid the town. They know there's nothing for them here.' Sadly, no mayors from other Romanian towns have ever asked him for advice – not even the mayor of Bușteni.

CHAPTER 17

A WAVE OF MADNESS

Even if the people of the plains sometimes pass over the mountain ridges, they pass like a wave of madness.

Bucura Dumbravă (1868–1926), Romanian author

Cluj, Romania's second city, has 400,000 inhabitants and is the undeclared capital of Transylvania. I've been coming here for more than thirty years, since communist times. It's an endearing mixture of the place I knew then, with over-sweet cakes, bottled fruit and peeling walls, and an imposing modern city. Much of the town is concentrated on the left bank of the Someş Mic, the Small Someş river, a left tributary of the Tisza, which rises in the Apuseni mountains, while the Someş Mare (Great Someş) rises in the Rodna range.

The cemetery, the botanical garden and the main square give the city its own distinct identity. Cluj boasts a strong football team, the annual Transylvanian International Film Festival (TIFF), and a range of ugly new shopping malls. The 'Napoca' officially attached to the name is a communist-era accretion that references what the settlement was called in Roman times – a political decision to downplay its Hungarian and German past.

The bronze Matthias Rex monument, of the king on his charger, stands in the centre of the city, beside St Michael's Church. My friend

Rachel de Candole says the statue is an example of the mistreatment of animals: it's clear from the horse's mouth that iron nails are being forced upwards inside, causing the animal immense pain – an impression that is reinforced by the position of the reins and the stirrups, and of the king in the saddle. This was a strange decision by the sculptor, János Fadrusz, she says, as his other equestrian statues show a great sensitivity to the well-being of the horse. The statue was unveiled in 1902, at the height of Hungarian power, as the country was emerging from the shadow of Austrian rule, anxious to impress the world. The statue is an exercise in macho sculpture.

King Matthias (r. 1458–90) was the 'good king' in Hungarian history, who stood up to the Turks, forced the rebellious barons into line, and established the Black Army, one of the first professional armies in Europe. His birthplace, a street away from the main square, is a handsome, white, two-storey building with a broad gateway and a host of chimneys, built in the fifteenth century and home to a wine-maker, Jakab Méhffi, whose name means 'Son of the Bee' or 'Son of the Womb' – the words for 'womb' and 'bee' being the same in Hungarian.

Folktales describe King Matthias dressing up in disguise and asking his subjects about their everyday lives, whether local officials were mistreating them and what they really thought about their king. The portrait of him on the blue, 1,000 Hungarian forint note is surprisingly ugly, with prominent nose, gluttonous lips, fat chin and hair reaching his shoulders. There is none of the humour or compassion the stories relate.

One story is called 'King Matthias goes a-thieving'. The young king sets out in disguise to visit a fortune-teller. She inspects the lines of his palms, then foretells a great future for him. He might even become king, if only he can survive his next name-day, she tells him. He will only avoid death on that day if he steals something.

King Matthias goes on his way, deep in thought. He has never stolen anything in his life, and doesn't know where to start. Then he hits on a plan. Disguised in poor man's clothes, he seeks out the dodgiest, darkest pub he can find. After a few drinks, a wretch in the corner confides in him that he has a plan to get rich the very next day, and if he will become his accomplice, he can share the proceeds.

The plan is a bold one. Leaving Matthias hiding in a bush, he produces a ladder from the undergrowth and climbs into the prime minister's garden and up to his window. But no sooner has he tipped the best silverware into his sack, than he hears voices. He hides under the bed. Lying there, petrified, he overhears the minister tell his wife to mix a strong poison into the golden goblet, which the minister plans to present to the king as a gift to celebrate his name-day. The wife protests, but the minister threatens her. If the king dies, he – the prime minister – will be proclaimed king and she will be queen.

When they are fast asleep, the thief climbs back down from the window. He confides the story to the waiting Matthias, and with it his sadness, because he loves and respects his king. But if he tells anyone in an effort to save the king's life, either no one will believe him, or he will be revealed as a thief and will face the gallows. Together, they bury the treasure and agree to meet the next day.

Matthias goes home to his own bed, and the next day greets the well-wishers at his name-day feast. Everything happens as predicted. But the king, grinning broadly, orders the hapless prime minister to drink from the golden goblet. When he tries to avoid this, the king accuses him of treachery and forces him, at sword point, to drink the poison instead of being hanged. Then the king summons the thief, who is richly rewarded – not just with a generous stipend for saving the king's life, but also with the hand in marriage of the newly widowed wife of the deceased prime minister.

By the end of Matthias's reign, Hungary was a European superpower – but one wrecked so swiftly and effectively by his successors that Hungary fell to the Ottoman Sultan Sulejman the Magnificent at the battle of Mohács in 1526, and remained a divided, occupied country for the next 170 years.

In a café just off the main square, I meet Réka Jakabházi. She's a literary historian, specialising in the poets – Romanian, Hungarian and Saxon – of the interwar years. Réka has reddish hair and wears a gingko-leaf pendant, and is delighted in my interest in the high mountain meadows of her research – including 'Transylvanianism', or the sense of Transylvania as a special place, set apart from both Hungary and Romania.

Each summer from 1926 onwards, Transylvanian writers (known collectively as the Transylvanian Helikon) gathered at János Kemény's castle at Marosvécs (Brâncoveneşti). Their credo was written by the Hungarian architect and artist Károly Kós:

> The Transylvanian Hungarians were different from the Hungarians of Hungary, the Saxons ... did not resemble the Germans of Germany and the Transylvanian Romanians were not like those living in Romania; they were all different physically, and even more so in mentality.

They had a 'shared destiny and common struggle with life', suggested Kós. For once, here was a writer reaching out to other cultures, exploring how their shared landscape made them alike, rather than searching for national differences.

In the German-language poetry of the interwar years, Réka explains, the mountains were portrayed as a place of refuge, where a Transylvanian identity could be forged. The stag, for all the poets, regardless of nationality, was depicted as a symbol of strength, beauty and innocence, beyond what Sándor Reményik called 'the plots hatched in Bucharest'. Lajos Áprily wrote of the 'mournful Szekler fir tree', defiant and alone on a desolate mountainside.

After years of mourning the loss of Transylvania, the Hungarian poets of Transylvania woke up one morning to find it was still there. The borders might have changed, but the landscape, the solitary fir trees, the outline of the mountains – all that was still the same. Away from the machinations of world politics, of Woodrow Wilson's plans and peace treaties, of the arrogance and illusions that led to the loss of so much territory, their sense of belonging, of geographical place, remained unbroken. The Hungarians still had a home, and a homeland, whatever anyone said to the contrary.

Hikers' associations regrouped after the First World War, to encourage people to explore the mountains, and to teach respect and care for them. Many of the hikers were women, and one of the most celebrated was Ştefania 'Fanny' Szekulics. Born in Pressburg (Bratislava) in 1868 into a Hungarian–German–Romanian family, she adopted the pen-name Bucura Dumbravă. She befriended the royal family of Romania and became a confidante and companion of the Romanian Queen Elisabeth before and during the First World War.

Bucura wore tailor-made clothes and devoted a whole chapter of her book to 'sensible garments for hiking'. She was passionately romantic about the mountains and especially the shepherds. In her *Book of the Mountains*, in the chapter entitled 'Sheepfolds', she wrote:

> Even if the people of the plains sometimes pass over the mountain ridges, they pass like a wave of madness. The mad wave washes over, but the rocks, the woods, the waters and the barns are left intact. And when, after the snows melt, the song of the water intensifies, the forest turns green and the pasture blossoms, then around the sour sheepfolds the bells mix with the bleating of the ewes and the barking of the dogs, and the good old man speaks of sheep, wolves and bears, and sometimes he indulges with the madness of the world. No matter how much this world boasts of its rights and powers, he suggests, it exists only through the work of the farmer and his honourable and industrious brother, the ploughman.

The Transylvanian Romanian writers' experience of the interwar years was radically different from that of the minority Hungarians and

Saxons. Safe in their new national unity, their most lyrical work focused on the horrors of the Great War.

The 1965 film of Liviu Rebreanu's 1922 novel *The Forest of the Hanged* begins with the marching ranks of the Austro-Hungarian army in the First World War, bobbing like logs being carried down a mountain river.

'During World War I, the Romanian province of Transylvania was still within the boundaries of the Austro-Hungarian Empire,' reads the subtitle text. 'Consequently, many Romanians were mobilized in the Imperial army to fight their co-nationals.'

In the First World War, Romania stayed neutral for the first two years, then joined the Allies against the Central Powers in August 1916. Romanians from Transylvania fought in the Austro-Hungarian army alongside Hungarians, Czechs, Slovaks, Poles, Ruthenians and Croats, against Romanians from the pre-war kingdom. While the Austro-Hungarian war aim was to keep the empire together, the war aim of Romania was to wrest Transylvania – with its 2.8 million Romanians in a population of 5 million – from the Hungarians.

For the Allies, Romanian allegiance was useful, in order to cut rail communication and supplies between Germany and Turkey, and to get access to Romanian oil. A bright-yellow British propaganda poster, entitled 'Rumania's Day', published in *Punch* in 1916, shows the German Kaiser leaning over a map of the battlefronts, addressing King Ferdinand I of Romania.

'So you, too, are against me! Remember, Hindenburg fights on my side!' The Romanian king, in dark-blue full dress uniform, stands defiant, leaning on his sword, and replies: 'Yes, but freedom and justice fight on mine.' The exchange, and the fact that Romania eventually joined the Allies against Germany, was historically poignant. The Romanian royal family, installed in 1881 to consolidate the country as it broke free from the Ottoman empire, were Hohenzollerns, imported from Germany.

In the 1965 film, the soldiers are marched up a muddy hill to witness the hanging of a Czech deserter, Svoboda. Apostol Bologa, a Romanian officer who cast the deciding vote at the court martial which sentenced Svoboda to death, attends the execution. He falls into conversation with a war-weary officer standing nearby, who turns out to be Czech, a fellow national of the man due to be hanged.

Despite his enthusiasm for order and discipline, Apostol is deeply disturbed by the execution. He tries to justify it in a series of conversations.

'The guilt was evident,' he says. 'The interests of state are superior to man.'

'Nothing is superior to man,' says Johan Maria Müller, lifting his flute to play.

CHAPTER 18

TO THE STONE PUB

In the hitherto wild valleys of the Carpathians, a devastating war is underway . . . squads of bulldozers and excavators are advancing, engineers expert in blasting and finally pouring in concrete. The tracks of micro-hydropower plants are crushing and destroying the life of natural waterways.

Péter Lengyel, biologist, 2012

The 15.41 train from Cluj to Braşov leaves on time. It's mid-winter, and ragged traces of snow decorate the earth beneath spindly trees, beside petrol stations and factories. Indecipherable graffiti cover rusting railway wagons and abandoned, communist-era buildings. Romania looks at her wintry worst.

Four hours' journey further east through the mountains, the snow transforms the landscape into a child's fairytale, sparkling white in the hooded lights of the train. Stepping down into a snowdrift on the platform in Topliţa, I laugh aloud. There is something so funny about deep snow, so soft and humorous, like a joke played by an infant God on an over-serious world.

The funny side of things evaporates when I reach a restaurant near the station. The kitchen is closed. There are some local youths glued to a slot machine. A surly waitress serves freezing beer beneath a neon strip-light. Stale salty sticks in a plastic bag are the only item available for the visiting gourmet. The place is not even warm. It's minus ten outside, and the short walk has soaked through my trouser legs, but fortunately not my boots. Oh for a roaring fireplace and a steaming bowl of soup! For a candle on the table and soft lights on the walls! And for the company of fellow humans with a tale to tell or listen to.

I've booked a room in a modern *pensiune* which looked close on the map, but turns out to be half an hour's trek through the snow. It seems completely empty, but the door opens when I try the handle. It's warm and dry, the lights work, there's an empty cold drinks machine and table football, but no other sign of life. My key is on the reception desk, beside my name. It fits the lock on the door. I shower, sleep and set an alarm. At 06.30 the next morning I set out again, without having seen a soul.

I'm supposed to catch a Dany Trans minibus in front of Banca Transilvania, to continue my journey south. Bogdan, who works at the national park, is efficient and friendly, and texts me messages with precise instructions. Everything works like clockwork. I even manage to buy myself some sweet buns for breakfast in a supermarket beside the bus stop.

The minibus is on time, and the bus is full of Hungarian-speaking men and women going to work and children to school. After four stops I step down in a village, and Bogdan is waiting in his car to sweep me to the office of the Bicaz Gorges – Hăşmaş National Park, where its director, Barna Hegyi, is waiting for us with a steaming pot of coffee. It's still only eight o'clock.

Barna is the only ethnic Hungarian head of a national park in Romania. It's one of the smallest and wildest of Romania's twenty-four national parks, with nearly three quarters of its territory regarded as 'core', which means it should enjoy strict protection. There used to be illegal logging here, he says, but stricter regulation has helped reduce it. The new threat is off-road vehicles, buggies driven at speed in the most sensitive parts of the park, crushing rare lichens and flowers and disturbing animals and birds. He only has six wardens, each responsible for around 1,100 hectares. The area is too big to protect properly. Even if people are caught by the rangers or the police, the fines imposed by the courts are minuscule, and foreigners never pay them anyway.

The main road between Moldova and Transylvania crosses the middle of the park and the Bicaz gorge, where the Bicaz river carves a path through the limestone, to empty into the Bistrița, which flows into the Mureș. Romanians commuting to western Europe picnic close to the road, throw down their litter, take selfies of themselves with a view of the mountains, then continue their journeys.

The 'quality' tourists that the park would like to attract – hikers and kayakers, rock climbers and ornithologists – make up 0.5 per cent of the visitors, Barna explains, but they are trying to attract more.

Between 150 and 200 bears overwinter together in isolated limestone caves – a rare phenomenon for such a lonesome, unsociable creature. A bear village. Thanks to the abundance of bears and wolves, and even some rare lynx in the park, there is no problem of deer preventing the natural regeneration of trees. The large predators maintain the ecosystem. The bear cubs stay with their mothers for two years or more.

Barna has battled for years to prevent new ski resorts being built in the national park. One project wanted to bulldoze the top of a mountain to install a ski lift: 'The politicians come to us and say – "there's nothing there, what's the fuss?"' Barna produces a wildlife map of the park, compiled by his rangers, which shows the bears, wolves and rare plant sightings throughout the park. 'This normally shuts them up.'

'By all means ski when there's snow on the ground,' he told the planners, 'but carry your skis up the mountain first!' Less and less snow is falling each winter due to global warming: that may stop the ski resort more effectively than his environmental arguments. The investors, he hopes, will turn away from his green slopes.

In terms of blind development and corruption, the Hungarian political party, the UDMR, is worse than the Romanians, he says, setting aside his loyalty to his own community. They place articles to influence public opinion in their favour. He fights to prevent mini-hydro projects that destroy mountain rivers and generate precious little electricity.

Something good can come out of every bad situation: the huge fuss over hydro projects and potential ski slopes served as an advertisement for the beauties of the national park. 'Now at least more people come here to see them!'

Further downstream from the Bicaz gorge, Lacul Roșu – the Red Lake – is another gift of the river. It was formed by a landslide of clayey soil in 1837, then consolidated by an earthquake in January 1838 which blocked the river. The scientific explanation for the red colour is the iron oxide of the rocks in the bed; but local legend offers another version. A beautiful young woman 'her figure as slim as the mountain poplar', fell in love with a local lad, who was then conscripted into the army. While he was away, a bandit abducted her; but before he could rape her, 'the mountains came to the rescue'. An earthquake buried the whole mountainside, including the girl, the bandit 'and even a gentle shepherd, grazing his flock across the valley' beneath the rocks.

The lake that formed has been stained by their blood to this day. But on a summer's evening, just before sunset, if you gaze into the waters, you can see the girl's sad, grey-green eyes, looking back at you.

◎ ◎ ◎

It was in Budapest that I first met Feri Ségercz, a maker and player of all manner of flutes from the Szeklerland. 'Come in mid-winter,' he told me, 'it's the only time you'll catch people at home. The rest of the year they're out working in the fields.'

Feri buys simply carved flutes, then fine-tunes them to sell direct to musicians, or through specialised music shops. He's a Szekler, but also a man of the world, not hemmed in by his mountains. In one of his trios, he plays with a Syrian from Aleppo who got asylum in Romania.

We set out together in his red car, with his black dog Bonifác, for a three-day trip through the region. The Szeklers are a fiercely independent-minded tribe of the Hungarians, proud of their Hungarian identity, but also a step apart from them. Sharp-witted, and quick to anger and love:

> The Szeklers have brought their own characteristic contribution to Hungarian culture and literature. Obstinate, at times downright laconic and sullen, hot-blooded, cunning and reckless, they are characterised by the nickname *góbé*, encapsulating their devious reasoning, sharp-tongued and outspoken nature, special physical and linguistic environment, folklore, traditions and art and peculiar dialect.

So wrote Andrea Tompa, in an essay on the Szekler playwright Áron Tamási.

The Szeklerland occupies two counties of modern Romania, Harghita and Covasna, but also spreads into Mureş county, around the city of Târgu Mureş. The men are famous for their fighting prowess, and the women for their beauty and marvellous costumes. Their folktales often involve trickery and supernatural characters, and explain specific features of the local landscape.

The sources of the Olt and Mureş rivers are only 20km apart; the rivers flow in opposite directions and define the Szekler heartlands. The Mureş is nearly 800km long, steering north and then east, to enter the Tisza near Szeged on the Great Plain of southern Hungary. The Olt is 600km long, flows south, then curves briefly north again, only to cut south again across the plateau of Transylvania, to carve a surprise path between the Făgăraş and the Retezat mountains, to enter the Danube near Turnu Măgurele. The waters of both end up in the Black Sea.

As we drive north, Feri describes the *kaval*, the strange flute that will accompany this whole journey. It's longer than the shepherd's flute, is often made from plum wood, and has between eight and five holes, depending on the region: those Feri makes have six holes. The instrument probably originated in Turkey, and entered Transylvania through

Moldavia. Economic migrants, Romanian speakers, were attracted by the good soil in the mountain valleys and the abundance of water, but brought their own music with them.

The kavals of Romania, Bulgaria and Turkey are very different. The Bulgarian has eight holes plus one at the back for the thumb; its wood is dark and very beautiful. Our conversation meanders easily between music, wood and water.

'Here, the water flows out of the wall,' he quotes a local saying. There is a double meaning: the sheer number of springs in these valleys, and an era before mains water and a tap in every home. Streams gush through the villages of his childhood, the water held by the land. That is all changing now. The cutting of the forests and the regulation of the watercourses mean it all flows away; the land has lost the ways it used to keep it.

We drive down crisp, frosty roads, past houses with brown, tiled roofs, the mountains in the distance lightly dusted with snow. Peasants sell string sacks of potatoes and onions at the roadside. The hills ahead are cone-shaped, volcanic; the fields are flat and ploughed, the grasses yellowed from thawed snow. Bonifác the dog perches on the back seat in the middle, and pokes his nose between us eagerly. 'He loves to travel,' Feri says, using the Hungarian phrase *nagy tekergő*, which could be translated as 'he's a great rambler'.

Bonifác is a *puli* – a Hungarian breed with long, woolly hair, famous for its loyalty and hunting skills, with the temperament of a black labrador.

'The communists poisoned folk music here with "Moiseyev" music [a kind of popular, sentimental dance-house music named after the Soviet choreographer Igor Moiseyev, much favoured by communist regimes in eastern Europe] . . . 'The dance-house revival of the 1970s was a reaction against that.'

Each era, it seems, has its sentimental garbage. The Hungarian 'popular' (*népszínmű*) theatre of the nineteenth century, according to Andrea Tompa, was a vaudeville-type genre, built on romanticised images, [and] portrayed a false, stylised peasant world which had never existed, in which the characters appeared in velvets and silks, sang imitation folksongs and engaged in complex amorous relationships across social divides.

We drive through Csíkszentsimon (Sânsimion), home to the legendary Csíki brewery. 'Real Csíki Beer' was the brainchild of András Lénárd, a Hungarian businessman with strong political connections to the Orbán government and a dodgy criminal record. For a while in the mid-2010s, he emerged as an unlikely folk hero, battling a giant multinational brewery. But there was more to the story than met the eye.

Romanians are among the biggest consumers of beer in the world. When capitalism returned to Romania in the 1990s, big foreign breweries moved in swiftly. By 2021, annual per capita consumption was 95.6 litres, just ahead of the beer-drinking Germans and Poles, though far behind the thirsty Czechs. In Miercurea Ciuc, the Dutch brewer Heineken bought the local brewery to produce Ciuc beer. Ciuc is the Romanian name for the town, and the brew was marketed in Hungarian as 'Csíki' beer (pronounced 'cheeky'). In 2013, András Lénárd popped up in a converted distillery in the village of Szentsimon with a small craft brewery, producing what he called 'Real Csíki Beer'. Heineken sued him.

The Romanian courts initially ruled in his favour, but Heineken won in the Appeals Court in Târgu Mureş in January 2017, effectively banning 'Real Csíki Beer'. András Lénárd and his publicity team skilfully wove this apparent setback into an advertising campaign, presenting their product as 'Forbidden Csíki Beer'. A black-leather punchbag was set up, marked 'Heineken', which visitors to his brewery were invited to hit as they passed.

High officials from the Orbán government in Budapest arrived to support him: the story was a perfect opportunity for the government to stand up for Hungarian minorities abroad. Earlier, in 2010, when

Fidesz came to power, they had replaced the EU flag in front of parlia-ment with the eggshell-blue Szekler flag, as a gesture of support for Szekler autonomy in Romania.

The Hungarian government threatened to ban Heineken in Hungary, on account of the red star on its cans and bottles – forbidden as a symbol of dictatorship (along with the swastika). Sensing that they were in a battle which they might win in court, but lose in the hearts and bellies of the beer-drinking public, Heineken opened negotiations with Lénárd, and in May 2017 they dropped all legal action. The two breweries could exist side by side. As John-Paul Schuirink, director for global communication at Heineken, told the *New York Times*: 'We recognized the emotional value of the Csíki brand name to its brewers and consumers, as well as to its stakeholders in both Romania and Hungary.'

That was the tale served up to a public hungry for David and Goliath stories, but there is another version.

According to an investigation by the Cluj-based online media outlet PressOne, Lénárd was prosecuted in the US for trafficking foreign girls across the border to work in striptease bars in Canada. He also had strong international financial backing, which rather undermines his underdog story. The money with which he set up the Csíki brewery came from the overpriced sale of six environmentally harmful mini hydroelectric projects on Transylvanian rivers, to Hungary's energy monopoly MVM. MVM paid €27 million for the controversial, unbuilt hydro projects on the Uz, Başca and Bărzăuţa streams, all in Natura 2000 protected areas. Even before this, he collected money for the project from a mysterious Cyprus-based offshore company, run by another Transylvanian Hungarian, and from the pension fund of the Hungarian Reformed (Protestant) Church.

An animated video produced by Csíki beer shows a mountain stream with a Csíki beer bottle wedged among the rocks. A young Szekler play-fully squares off with a huge bear, but the bear points out another man: 'Watch out, someone's trying to steal our beer!' 'And he's wearing clogs,' says the thief, whereupon the bear biffs him. In other words, he's from the Netherlands – a reference to Heineken. The film ends with the bear and the Szekler ambling off together with their beer through the woods, and the message: 'The Bear is with Us.'

The 'Forbidden' label has been used as a marketing ploy by Csíki beer ever since.

'For each product the story is crucial,' Lénárd told the *New York Times*. '"Forbidden" will remind people of the story. Like in whiskies, nobody smuggles Old Smuggler but it's good to have a hint to the story in the name.'

There were strong nationalist undertones from the start: the Hungarian backing for Lénárd, the government use of the story in its own propaganda, its attack on Heineken and the swift arrival of Orban's chancellor to raise a glass with Lénárd. The afterlife of Csíki beer as a symbol of the Szekler campaign for political autonomy suggests this was all brewed up in a kitchen in Budapest, rather than Szentsimon. Belatedly, the Romanian state got involved, trying to ban the spin-off Csíki crisps in June 2022 for writing 'Szeklerland' as the country of origin, rather than Romania. For what it's worth, Csíki beer tastes a lot better than Ciuc and is popular in both Romania and Hungary. Free publicity has boosted sales.

The short winter's day is ebbing fast, the sunlight already catching the steep sloping church roofs, and hurrying up their towers. The village council buildings fly three flags here: Romanian, Szekler and EU. The Szekler flag is rather beautiful, a golden sun and sickle moon on an eggshell-blue background. It's a new design, based on ancient symbols, such as those found on the great carved gates of the Szeklers. The European Union flag – a circle of stars on a dark-blue background – was inspired by the circle of twelve stars around the head of the miraculous statue of the Virgin Mary in the pilgrimage church at Csíksomlyó, only a few kilometres away.

We pass through Ditró (Ditrău), scene of an ugly incident in 2016. The local bakery, which employs eighty people and serves a whole network of villages and towns on the Csík plain, could not find enough workers. Jobs were offered to two young Sri Lankans. Influenced by the constant stream of hostility to 'migrants' in Hungarian media, 'several loud-mouthed locals', as Feri puts it, led by the Catholic priest, decided to drive the dark-skinned Sri Lankans out of town: 'If someone learns a profession here, let's say as a baker, then the next day they go abroad to work, because the pay is much better.'

As Feri explains, the conflict wasn't only – or even mainly – about race, but rather about the low wages offered by the mean bakery owner. That created conflict between the bakery and the local people. His decision to employ Sri Lankans was the last straw. The incident did enormous damage to the good name of Hungarians at home and abroad.

We cross the Maros (Mureş), which will keep us company for the rest of the journey, and stop in Gyergyóremete (Remetea) to deliver a batch of Feri's flutes. A group of eager children, their parents and folk-dance teacher Lajos Rigmányi, wait for us in the village cultural centre. The room is like an old-fashioned cinema, with folding chairs and heavy velvet curtains.

Soon there's a marvellous cacophony of children blowing as hard as they can. I watch Lajos carefully, but there's no annoyance, no stern intervention. Like any good teacher, he wants them to have fun first, before the real teaching begins.

Feri gives a demonstration, sitting on the edge of the stage, Afghan hat perched on his head, flute tilted to one side. The children fall silent to listen. How easy it must have been for the Pied Piper of Hamelin to lead the children away. There's something hypnotic about the music of a small flute, which gives you no choice but to follow.

'It ought to sound,' Feri explains to the giggling children, 'as though you've got a wasp trapped in your throat.'

When we emerge, it's almost dark and snow is falling by the shovelful. The shore of the Maros bristles with thermal waters and baths, including the famous *borvíz*, the naturally sparkling mineral waters of the region. Such riches are hidden in the blizzard through which we skid, threatened by giant trucks lurching into view, before disappearing in the gloom of the Görgényi (Gurghiu) mountains to the left of the road, and the Călimani mountains on our right.

We retrace by road the route I travelled a few days earlier by rail. There are little blue Szekler flags not just on the town halls here, but on the houses. A fox stands uncertainly beside the road in the snow, its fur lit up bright orange in the streetlights.

It's too late and too snowy to visit the flute makers now, so we press on to Târgu Mureş, where Feri's girlfriend is waiting for us with a hot meal and a selection of excellent local craft beers.

The next day, we visit Hodac, a village to the north of the city. In March 1990, three months after the revolution, Romanians from Hodac fought ethnic Hungarians on the streets of Târgu Mureş. Some feared the outbreak of civil war, but common sense prevailed, and Transylvania avoided the fate which befell Croatia and Bosnia.

I warm my hands and knees in front of a wood-burning stove in Nicolae Feher's workshop, and discuss music, the composer Béla Bartók and the events of March 1990 in the same breath. Nicolae is embarrassed that people from his village took part in violence against Hungarians. 'It certainly wasn't us, or anyone from our families,' he told me.

Feri refuses to take a Hungarian nationalist line, which blames all the events of those days on the Romanians. He has always been treated with respect by his Romanian friends in Hodac, he says, and the musical genius of the two peoples is closely intertwined.

These kavals are made from hazel, cut on the edge of the village. They produce a deeply satisfying sound, very different from the high pitch of the shepherd's flutes. Then we cross the road to another flute maker. Feri brings back a bag of kavals he's not satisfied with, to be repaired. The man takes this news in his stride, and agrees to carry out the work. Feri blames himself. 'Either I cut them too small, or they got worn away.'

In my audio recording of the man's workshop, there's the sound of him sanding down flutes; Feri trying them out one by one, like a bird singing on a branch; and under it all, the steady murmur of a pot of meat stew bubbling on an iron stove.

Three months before the First World War, in April 1914, the Hungarian composer Béla Bartók set out with his wife Márta through Transylvania on a trip to collect folk melodies. It would be Bartók's last chance to tap the musical genius of a village world about to be turned upside down and lost forever.

His month-long trip through the valley of the Upper Mureș was particularly fruitful. Until then, he had mostly collected melodies in majority ethnic Hungarian or Romanian areas. The beauty of the Upper Mureș was the intermixing of the musical styles of the two peoples and the impact of music from Moldova and Bukovina, drifting across the valleys. A perfect setting for inter-ethnic musical exchange and tolerance. It was also the region which had sparked his original interest in folk music. At his home in Budapest, Bartók had overheard a maid from Chibed in the Mureș valley singing. Astonished by the melodies, he set off to find out where they came from. One of the villages he visited that April was Idicel.

It has stopped snowing by the time we arrive there, but the drifts are so deep that we leave the car as far up the hillside as it will safely go, then walk the last half hour, Bonifác leaping ahead through the snow.

Ion Covrig, a shepherd and musician in his early eighties, lives in the last house, at the end of the track. He serves plum brandy as a welcome drink. Except that it's half-plum and half-pear; he apologises – a bear ate half the plums from his trees last September. It tastes just as good. Ion has a craggy round face and deep blue eyes, and looks at least ten years younger than his age.

A small chicken flutters round our feet. 'It was a bit poorly, so we brought it inside . . . then it got used to the warmth, so we couldn't turn it out again,' his wife explains.

She serves plates of steaming polenta, topped with home-made sheep's cheese and sour cream, and sausages for the meat eaters. This is washed down with more brandy and the juice of red cabbage, a delicious, salty leftover from the fermentation process.

When we've eaten our fill, Ion lifts a kaval to his lips. The wood is dark with age, the cracks in it repaired with what looks like fine muslin, but turns out to be pig's bladder. He plays a shepherd's tune designed to carry far over the hillsides and draw the sheep back to the enclosure at dusk, when the bears and wolves begin to prowl. Then a round dance, devised to get the boys to their feet and the girls into their arms. He pauses after only two tunes and hands over to Feri, dissatisfied with his playing. One of his teeth fell out recently, he apologises, and the side of his mouth hurts.

They have five children, including a daughter who lives in England. I ask how she copes with the English rain, and get a novel answer. She likes it, her mother says – it's warm! How cold the rain must be sometimes, by contrast, up here in the mountains above the Mureş! She even has a boss who is Chinese – or possibly African, her mother can't quite remember. She and her family were home over Christmas for a few weeks, and have only just gone back.

The wind whistles impatiently at the window, and the cat plays with the chicken on the kitchen floor. We leave reluctantly, weighed down with big plastic bottles of milk from their cow, and stride through the snow in search of the car.

To keep him awake at the wheel on our long journey, I tell Feri my own tenuous link to Béla Bartók. In July 1988, in the Farkasréti cemetery in Budapest, I reported on his reburial.

Bartók died in New York in September 1945, at the age of sixty-four. During the fading days of communism, the authorities and his family struck an agreement to allow his mortal remains to be reburied in his native soil. The most surprising moment of the funeral was seeing someone who looked just like Bartók himself, at his own graveside. It was Péter, his son from his second marriage.

Down in the Mureş valley we stop at Marosvécs to see the castle where the writers of the Transylvanian Helikon gathered in the 1920s. We arrive unannounced and the castle is deserted.

Géza Kemény invites us to his handsome, but not quite so imposing, home in the next village, Galonya (Gălăoaia). Approaching up a snowy road, the building is distinguished by a single tower. Inside his living room, black-and-white drawings of the writers line the walls, beneath photographs of the natural beauties of the region, and the odd Hungarian and Szekler flag.

Sándor Reményik looks fierce and moustachioed, his white shirt open at the collar. János Kemény – the owner of the castle in the 1920s and Géza's ancestor – looks tidier, beside his wife Augusta Paton. Born on a Greek island, Augusta first visited Transylvania on the back of her brother John's motorbike in 1922. John, a farmer from Scotland, found himself in Transylvania when war broke out, researching the possibility of importing the famous Hungarian *mangalica* pig to Scotland.

Initially interned as an enemy national, the combined efforts of the Transylvanian Hungarian aristocracy ensured his release, and he spent a pleasant war learning Hungarian with his hosts in one stately home after another. On their visit in the summer of 1922, John's motorbike, bought from a demobbed American soldier, broke down at Marosvécs. János Kemény took such good care of Augusta while the motorbike was being repaired, that they fell in love. They had six children, five of whom survived, and split their lives between Scotland, Greece and Transylvania. Though the communists took away all their properties in Romania, they continued to visit, to the end of their long lives, in the 1980s.

Next to Augusta and János is the portrait of another poet, Jenő Dsida. Almost alone of the Hungarians, he managed to avoid politics in his writings. In the portrait, the fine, thin lines of his face, his neatly combed hair, the ring on his left hand give him an air of great intensity. His most beautiful poems were addressed to his wife, Imelda:

and your sigh is the wind
stirring in my hair,
and your face has the moon's glow,
and your face the sun's glare.

In the photograph taken in front of the castle gate at the Marosvécs gathering in 1926, three women were present among twenty-one men – Mária Berde, Irén Gulácsy, and János's sister.

Lajos Áprily left Transylvania for Budapest in 1929, to the dismay of other Hungarian writers, who accused him of betrayal. The new secretary of the Helikon, Aladár Kuncz, invited Mária Berde to write an essay on the difference between 'to believe' and 'to commit' – in Hungarian, almost the same word, differentiated only by an accent and a lone 'l': *vallani* and *vállalni*.

'Must the modern Transylvanian writer directly address current painful and topical issues,' Kuncz asked, or was it acceptable to 'bury oneself instead in the flowery language of history'? It was a double-edged question – about the literary nature of the historical novel, and the political responsibility of the writer to his or her community.

Mária Berde's answer was clear. Historical novels which reach from the past into the present with 'sharp and lively connections' were acceptable, she wrote. But not those which try to 'smuggle our present, separate truths back into the past' and try to 'artificially inject the present into the past'.

'From a purely literary point of view,' she added, 'it would not only be more honest, but also in some way easier to capture the life that pulsates and offers itself in the present.'

Her article provoked a storm among Hungarian writers. 'The majority of Transylvanian writers have sold their independence,' wrote Gábor Gaál, editor in chief of the *Korunk* (Our Era) periodical, 'for an incompetent uncle who installs inhibitions on the writer's emotions' – a reference to Count Miklós Bánffy.

Gaál accused the Helikon writers of acting as apologists for the injustices of the past, and demanded a more earthy, committed realism:

> This secretive academy invented [the idea] that the Székely peasant could be reconciled with feudalism, that the urban citizen could be reconciled with it, and even that . . . the gentry's now out-of-date consciousness can do it.

Weary of the bitterness of the debate that her essay ignited, and of the hostility between the men in the group and those outside, Mária Berde quit the Helikon circle and went back to teaching.

After a *pálinka* or two, Feri and I take leave of Géza Kemény's generous hospitality and drive back along the valley of the Mureş. Our destination this time is Miklósvár (Micloşoara), home to another Hungarian count who has reclaimed some of his ancestors' properties, Tibor Kálnoky.

The Kálnokys are well known in Romania for hosting Britain's King Charles. They own houses spread in a triangle of villages in a curve of the Olt river, Miklósvár, Zalánpatak (Zălanului) and Kőröspatak (Valea Crisului).

Tibor helped Charles find and renovate a property in Zalánpatak, and introduced him to the beauties of Transylvanian architecture and landscape. He even set up a joke 'throne' for Charles: an armchair by the window, overlooking the farmyard. The future king was delighted.

Once I'm settled in my own luxurious quarters – the Hussar's Room – Tibor shows me round. It's hard to describe Kálnoky's Transylvanian guesthouses without sounding like a travel guide. The ceiling of my room is of ancient timbers, with a magnificent tiled stove emitting a steady and drowsy heat, an idiosyncratic bathroom, comfortable chairs, old paintings, solid oak furniture, painted chests, traditional woven rugs and carpets, and lace curtains on the windows. There's a hidden heating system, so that guests don't need to learn the trick of keeping a wood-burning stove alight all night.

In the evening, over dinner we sit at a big table with Tibor's father and discuss Donald Trump, the twists and turns of global politics, climate change and the future king of Great Britain and Northern Ireland – all washed down with Romanian and Hungarian wines.

The Kálnokys can document their ancestors all the way back to 1252. The Szeklers were traditionally border guards. Miklósvár was first mentioned in 1211 as a border settlement between the newly arrived Szeklers and the Saxons.

'There were always problems here with Turks, Mongols, Tatars, Russians, Germans and Habsburgs. Everybody loved to loot and burn, which has made it more difficult to keep track of things.'

A dozen people work full time for the estate. He funds his business with the profits from the guesthouses, and uses grants and loans to develop the properties.

'We're living in a healthy community. Everybody sees everything, and they take care of each other.' All the villagers are Hungarian, apart from two Romanians who bought weekend houses, and one Saxon.

Tibor grew up in western Europe, mostly Germany, before coming back to Romania after the revolution to claim family properties seized in 1948. He describes the process as both a crazy dream and an ordeal. He's done better than most: more than thirty years after the fall of communism, huge numbers of claims remain unsettled across the country, and some decisions on church and school properties have been rescinded.

In the grounds, there's a small, astonishingly blue – almost purple – house, a variation on the Voroneţ blue I saw in Bukovina. A colour the Romanians brought to Transylvania, Tibor says, the colour for serfs' houses and stable buildings. The Hungarians in the village use green or white, so they looked at him askance when he chose that one. He loves it as a stunning colour, especially striking today with the snow all around.

We drink tea in front of a roaring fire in the Stone Pub. Feri plays to the assembled company, the rolling lilt of his kaval like the sound of a horse and cart rumbling down a country lane. Then the tune speeds up, and I can imagine the cart arriving on market day in a bustling square, pushing gently through the chattering crowds before tying up beside a fortified church.

Raluca is another visitor: she makes medicinal herbal teas at her home in the foothills of the Călimani mountains. Her tea is from calendula, ribwort, mint and sage, grown in her garden, mixed with blackberry leaves from the mountains. This one is called 'Good for everything' tea, and is distinguished by the bright orange of the calendula flowers. The teas are a spin-off from her company Hodaia, which makes cosmetics from oils and elixirs. The word *hodaie* means a room or homestead, a sheepfold or village hearth. When Romanian villagers from the Transylvanian hills say they're going *hodaie*, it means they're going home. Tibor puts another log on the fire and pours a *pálinka* to complement the tea.

The names of the carpenters are carved on the beams over our heads. The pub is mostly for his guests and friends – villagers who need a drink mostly go to the Miki Kocsma, a bar 'which hasn't changed much since the 1950s'. Eastern European villages are haunted by glum, hopeless places where people go to get drunk and stare into their glasses, rather than to enjoy themselves. There is no equivalent of the English pub, which makes Tibor's attempt almost unique. But for the pub to take off, he would need to employ an innkeeper. 'And do I really want to attract drunk people, who come round looking for an argument?'

The castle has been renovated with a grant from the Norwegian government. One condition is that he cannot make a profit from it. Tibor and his wife Anna spend the weekends at the family home in Kőröspatak and weekdays in Miklósvár, overseeing the guests and the renovations.

Originally, Prince Charles' house in Zalánpatak had only three rooms; but as he normally arrives with a team of a dozen people, including bodyguards, the house has had to be expanded, while other buildings in the tiny village have been done up to accommodate them.

The wildflower meadows of the region are Charles' passion. Twelve types of orchid can be found in Zalánpatak alone. He tried to establish a Transylvanian garden at his home at Highgrove, Gloucestershire. For years, Tibor sent his royal friend the seeds he collected himself in the meadows, and even bags of Transylvanian soil; but all to no avail. Charles wrote plaintive reports recording the appearance of 'a single orchid' or 'a few of this or that'. Gloucestershire was no match for the Szeklerland.

Tibor finds it hard to pinpoint where he feels most at home:

Kőröspatak is the family's real home. But I like the crypt we have here much better than the one there, so maybe my final resting place will be here, with a beautiful view over the valley.

Before the evening concert in the cellar of the castle, Tibor and Anna show me round the building above. A long corridor where more rooms are being prepared for guests; a green tiled stove, reconstructed by master stove-builders from fragments found on the floors; paintings of Tibor's ancestors on the walls, including one of Sámuel Kálnoky. A former governor of Transylvania, he fought on the Habsburg side in the battle of Zărneşti in 1690, against the Ottoman-backed Transylvanian Prince Imre Thököly. In the portrait, he wears a bright-red tunic and has long hair and curling moustache; he doesn't look much like Tibor, except for his dark eyes and his smile.

The concert begins at six. As we're settling into our seats, I ask Feri about the difference between folk music and classical music.

'Today's classical music is tied down to a precise musical score, while folk music lives from its variations,' he explains. 'The folk musician constantly improvises on a melody.'

The previous evening over dinner, he had stunned the Romanian musicians with his knowledge not only of Bartók, but also of the great Romanian song collector and composer George Enescu. After several days on a diet of purely folk music, I'm looking forward to what a pianist and violinist can produce in the bowels of a medieval castle.

Violinist Diana Jipa and pianist Ştefan Doniga are on the last leg of their 'Romania Universalis' tour – 'music which celebrates the diversity of Romanian culture by approaching a selection of pieces composed by Romanian composers with several ethnic origins'. I wonder what Bartók would have thought about being introduced as a Romanian composer.

The concert is enchanting, like a ride in a luxury car through the villages of Romania. Enescu was a child prodigy who died in penury in Paris. Later, in the Stone Pub, Feri plays a melody from the village of Nyárádmente (Valea Nirajului), south of Târgu Mureş. There cannot be many places left in the world where you can place a tune or a song to a specific village.

CHAPTER 19

THE GYPSIES OF THE ŐRKŐ

By blood, I am Albanian. By citizenship, an Indian. By faith, I am
a Catholic nun. As to my calling, I belong to the world. As to my
heart, I belong entirely to the Heart of Jesus.

Mother Teresa of Calcutta (1910–97)

In May 2014, I spent a week in Sepsiszentgyörgy (Sfântu Gheorghe),
filming with the Gypsy – or Roma – community in the Őrkő district.
The Gypsies are all Hungarian speaking.

András Márkus, the Catholic priest, is a jovial man with the patience
of Job. After a visit by Mother Teresa to Romania in May 1990, the
Church decided that something had to be done to help the long-
suffering Roma population. Father András was sent by the bishop in
1991. He describes walking round with a picture-Bible in his hand,
asking children to read the text. None could. Around 2,000 Gypsies
live here, almost completely segregated from the Hungarian majority in
a town of 60,000. They used to work in the stone quarry or the brick
factory, but both closed down after the fall of communism.

First a new school was built, then a new church. András was helped
by Mother Teresa's order – the Missionaries of Charity – from India, the
original homeland of the Gypsies. This helped relations between

Church and community – clean, chaste, prayerful Gypsies in blue-and-white robes, setting themselves up as role models for the wild Gypsy girls.

Wherever we go in the narrow, higgledy-piggledy streets, a small crowd of ragamuffins gather in front of our camera, hoping for a copy of the 'photograph'. They stare straight into the lens and pull faces. This becomes the trademark style of our film. The older Roma are more wary, fed up with being counted and questioned, with no visible improvement to their plight. Families welcome us into their homes, and tell us their stories. Everyone complains about a new law, banning horses and carts from the main roads of Romania, and even from the main streets of this town. The law is a disaster for a people that makes a living collecting scrap metal, door to door.

I take their anger to the mayor, András Árpad Antal. He's been in power here since 2008, but seems unsympathetic to their plight. 'That's progress,' he says bluntly. 'The horses and carts were causing accidents, especially at night, without lights. They were a hazard to cars and other traffic. They ride their horses and carts the wrong way down one-way streets!'

Asked what message I should take back to the Gypsy ghetto, he answers: 'Let them buy themselves cars.'

It doesn't seem a very realistic suggestion in a community where most children don't even have shoes.

There has been steady progress in the number of children who complete eight years of primary school, András the priest tells me. The girls are more problematic, since the families believe they should be married off at twelve or thirteen, as soon as they start to menstruate. This is typical of many traditional Roma communities in eastern Europe. Their husbands often refuse to let them attend school anymore. The secondary school only has one class, as few children are motivated to study. Father András describes a visit from a Hungarian bishop who was concerned about how few Roma children in Hungary attend university. András told him that he should be happy, considering how few Roma children in Romania even attend secondary school. He cites the case of an eighteen-year-old boy who finished school, but failed his school-leaving exams. He got a job at the waste-disposal company – the same job as his fellow Roma who had left school at thirteen.

'Where was the value of studying for years on end, when he could have been working and bringing home an income for his family?' András asks. In 2014, ten students entered the secondary school – a new record. So there is some progress.

One day I go with Dima and his many children to fetch water from a spring. The water trickles out of a crude pipe that sticks out of the hillside. Litter is scattered all around, but the water is cool and refreshing.

'Is it safe to drink?' I ask.

'Of course,' replies Dima, without a trace of a smile. 'You only get sick the first few times.' In the film, I nod wisely, as though he'd told me he was planning to bottle it and sell it to the sheikhs of Saudi Arabia. One of his daughters grins mischievously. It's one of my favourite scenes in the documentary.

As the day of the confirmation draws closer, the excitement in the village rises. Timea, aged thirteen, stands between her mother and grandmother in their living room, in an ornate white-laced blouse and apron over a red-striped white skirt, with a purple velvet waistcoat, her dark blonde, shoulder-length hair constantly being brushed.

They're short of mirrors. Necklaces, rings, earrings and all manner of brooches are extracted from a pink tin box, with a mirror in the lid for her to check herself. She chooses a necklace of silver discs. One of her little sisters gets very excited as a white frock is lowered over her head, turning her into a snowflake. Some of the items of clothing are family heirlooms, like the wedding suit her father is wearing; but most of the girls' clothes are borrowed for the day. Timea's parents have been together sixteen years, her mum tells me, proudly.

On the big day, the sun shines brightly and everyone spills out into the courtyard. The blue house has dark-blue-and-yellow stencilled flowers along the tops of the walls. There's a carnival atmosphere, music booming from a beatbox, children's voices almost as loud. I take a picture of the whole family before we set out for church. Timea wears amazing high-heeled boots, walking between her mother, in dark trouser suit and yellow blouse, and one of her younger sisters, over the hard earth track leading up the hill towards the church. She and the other girls carry brightly decorated baskets of red and yellow roses, with copious ribbons and silvery tinsel. Other families join the procession. Everyone turns out to watch the procession from their doorways.

The crowd assembles in the street in front of the church. Most of the girls are in folk costume like Timea, but one or two wear more modern garb. A girl in a fluorescent orange miniskirt with black waistband and bolero cardigan leads a line of others in white blouses and aprons with dark-red waistcoats their plaits tied with red ribbons. Other girls wear ankle-length, flowery Gypsy skirts of yellows and purples with flower patterns. There are silvery blouses, large moonlike earrings and small golden ones. The next big ceremony in the children's lives will probably be their weddings.

The forty children to be confirmed enter the church first, for prayers with Father András. The boys on the right facing the altar, the girls on the left. Then the Indian nuns in their blue-white habits open the doors, and a great throng of family members and friends rush inside, carrying bunches of flowers. There's a baby in a little white bonnet, men with short, neat moustaches, everyone in their best clothes. In the tumult, the nuns try to make sure everyone finds a place to sit or stand.

There's an appeal for silence, which softens the hubbub a little. Then a nun with a guitar starts playing and everyone starts singing, straight from the heart. Father András walks up the aisle, accompanied by a crimson-robed bishop. At one point in our film, the camera pans up from the attentive faces to a high-up painting of the Gypsy Saint Ceferino Giménez Malla, martyred in the Spanish Civil War, standing beside a white horse. A girl with a flower in her hair and a strong, clear voice reads the lesson.

As the congregation sings, the children come forward in turn. The bishop makes the sign of the cross in holy oil on Timea's forehead. Above the altar behind him, Mary is young, rosy-faced and beautiful, her skin as dark as that of the adoring cherubs around her, her hair crowned with flowers, her arms outspread, blessing her people. Under the pews, the smallest children play hide and seek, and the final chorus of 'Alleluia . . .' rises to heaven.

'They're a deeply religious people . . .' says Father András about his community, choosing his words carefully.

And their faith has always been a superstitious one. There is great respect for the figure of Mary. I often tell the Hungarian faithful how marvellous it is when they sit in the church, the children's faces glowing, how loudly and enthusiastically they pray. Sadly, among the Hungarian believers, and no doubt believers of other languages, it seems that way of prayer has been forgotten, that sense of joy whilst speaking to God.

In his dark habit, sitting on a wooden pew in the darkness of his church, Father András reminds me of the Trappist monk and poet Thomas Merton.

After the service we go back to the Őrkő for the party. A man in a scarlet shirt plays an accordion. A woman in a purple skirt and blue waistcoat, her hair tied up in a flowery headscarf, begins to dance, then turns in mock indignation to the crowd – 'how can I dance without a man?' A succession of men of all ages, excellent dancers, step forward, flicking their heels in the air, slapping their knees, throwing back their heads in the ecstasy of it all.

When I return in January 2022, Father András is away for a knee operation in Hungary. The nuns are keen to help, and ring Feri Monu. He takes me for a tour of the neighbourhood. I have brought a few DVDs of the film I made eight years earlier, and want to find out what has become of the people I met then.

The streets look much the same as in 2014: little makeshift shacks, some only one room, wattle and daub, and the lively buzz of a people who spend half their time outdoors, even on a chilly January afternoon. Babies crying and dogs barking and men and women shouting, loud music pumping out of beatboxes.

Feri and I sail through the middle of it. Aged sixty-five, he is one of the older members of the community and is deferred to wherever we go. 'May God grant you strength and good health,' he says to everyone he meets. To those who he knows have recently lost a loved one, he adds, 'May they have eternal rest in Heaven, and may they find everlasting peace, in the Lord Jesus's holy name.'

Though Feri is not a priest, he moves through the streets, bestowing blessings and attracting them from the older people. Some children tag along for a while. They ask Feri about me: 'Who's the uncle? Is he a priest?'

We head to his house. It's a cold afternoon, but the wood-burning stove is well stoked, and two of his seventeen grandchildren – Henrietta aged seven and Nico aged four – are playing in the living room. Nico attends the church kindergarten, the boy tells me proudly, and already knows 'all the letters and the numbers, and I can write them as well!' Feri also has three great-grandchildren, and will soon, in 2024, celebrate his golden wedding anniversary – fifty years of marriage.

Henrietta says she likes to read. Their parents – Feri's children – are away in Hungary, doing any jobs they can find, helping with the grape harvest in September and October, pruning the vines at this time of year. His son also has a diploma as a stone mason.

We taste his homemade plum *pálinka*. 'I like it strong, like my father,' he explains. 'I don't like the weak ones, they have no flavour.' This one is very strong – over 50 per cent ABV, he says – but that doesn't obscure the fine taste of the plums. His front room is clean and simply decorated, with pictures of the Virgin Mary and bright rugs on the wall. His son and daughter-in-law are due home anytime, and the children are excited.

We talk a little about politics and the state of the Romanian economy. He doesn't understand why wages and welfare rise at barely half the rate of inflation, nor why the powerful never consult the ordinary people, who suffer as a result of their policies. About a quarter of the Gypsies from the district are abroad, including Timea and her family, he reckons. Those with a skill are working, the others are begging.

We visit his friend János, who might have a DVD player. He doesn't; but he does have an old laptop with a separate drive that can read the disk – though without sound. We watch the film as János's wife Mónika brings strong coffee, cooked on the stove, and people crowd around, identifying some of the characters from eight years ago. Those abroad, one who has passed away, another who is ill.

The *pálinka* flows, and soon everyone is shouting at once. There's some dispute about whether the town council is as bad as it's always been, or if the mayor is now doing something for the Gypsies. Construction of a new housing estate is planned on the edge of the ghetto. People are worried about the future.

'When we were younger, our parents only got state benefits if we went to school,' says Feri. 'But in those days – before the Gypsy school was built – the teachers didn't want anything to do with us, because we were Gypsies, so we didn't learn much.'

It's getting dark outside, and I want to get back to the Ferdinand guesthouse before nightfall. Feri, even with his bad leg, walks me some of the way, so I don't get lost. He insists on buying me a bottle of beer at Gigi's general store on the edge of the district. We clink bottles, then I walk off down the hill, into the less chaotic, less welcoming world of the non-Gypsies, the *gadjos*.

CHAPTER 20

WOLF, LAMB, LEOPARD

I am the metal left behind
from worlds turned into ash by crimes
I am the echo, echo, echo
from bygone worlds and bygone times

                    Toma Arnăuțoiu (1926–59), Romanian resistance
                                                fighter and poet

The East Station in Budapest is broad-shouldered, poor and damp in the early December night. A great vault of a station, into which trains rumble, glide and groan like ghosts.

The 19.10 overnight train from Budapest to Brașov leaves on time, to the split second. I'm sharing a four-person sleeping compartment with a Romanian in sports gear with his son, maybe ten years old, and Helmut, an elderly German. Helmut left Romania in 1979 – one of thousands of Saxon Germans 'sold' by the dictator Nicolae Ceaușescu to the West German state, for a few thousand Deutschmarks apiece. On a trip through communist Transylvania in the 1980s, I travelled in a compartment with a young Saxon girl, who wept all the way. She was the last of her family to leave, she told me.

Helmut and his wife went first to Germany, then Spain, on account of Helmut's asthma. He needed the salt-sea air in order to breathe. They spent twenty-five years together on the coast north of Barcelona, until she passed away in August. He came back to Romania to bury her ashes in Timişu de Sus, a village on the mountain above Braşov. Now he is back again, to sort out all the paperwork, with a view to returning to live in Romania for good.

'I'm all alone in Spain,' he says, in his thick, old-fashioned German. He seems broken by the loss of his wife. They have no children, just a dog, which is being looked after by a neighbour. Above all, he fears for his breathing in Braşov, in the damp, polluted air of a sprawling city at the foot of the Carpathians. He wonders if he is coming home to suffocate. Perched on the edge of his bed, he paints a picture of desolation. Throughout our conversation, he refers to Braşov by its German name of Kronstadt. Even his pale-brown suitcase, I imagine, once belonged to his wife.

One by one, we make our beds. The boy and his father play on their phones for a while. Helmut and I are the last to turn in. He has the lower bunk, I'm on the upper. By then we are passing the Hungarian city of Békéscsaba. We are all fast asleep when the Hungarian border guards wake us to check our passports. And all asleep again when the Romanians check them again, an hour later.

The next thing I know, white light is leaking into our carriage, round the edge of the flimsy curtains. I look out onto a Transylvanian world white with hoar frost. Telegraph poles, half sunk into the ground, left to fall apart in a digital age, thin, icy cobwebs strung between their fairy cables. The water in the ditches is not frozen, so the cold cannot be too severe. The train stops in the village of Făgăraş.

As we pull into Braşov, Helmut's face grows darker. He takes a final blast on his inhaler. I help him down with his suitcase onto the platform, bid him farewell, then set off. When I look back he looks so forlorn, frozen motionless on the platform, I go back to help him. '*Vielleicht nur bis zu den Verwandten . . .*' he mutters. If I could just help him pull his suitcase as far as his relatives' car, that would be a great service.

We have to stop several times for him to catch his breath. Outside, a large, very smart car pulls up, and a sharply dressed man in a waistcoat and wide-brimmed hat leaps out, followed by a young woman of fashion, with bare arms and a lot of jewellery. I'm introduced to Mr Codreanu, owner of the main bus company in Braşov, and a former colleague of Helmut's from the 1970s. I'm happy to accept a lift into town – 'one good turn deserves another,' says Helmut, a bit more cheerful. They drop me near the Black Church, my main landmark in the city. It feels good to be back.

Braşov in December 2022 shelters beneath a massive Christmas tree in Piaţa Sfatului (Council Square), gaudily bedecked with gold and silver baubles. The Black Church only opens at ten o'clock, so I take refuge in a promising-looking café opposite the main entrance, and am not disappointed. Long wooden tables, full of families with small children, stray men with dogs, delicious coffee and a fresh pistachio cream-smothered croissant to quell my hunger. Overhead, the top floor has been removed to reveal a restored timber roof.

Dan, sitting opposite me across the table, is an alpine climber, not to be confused with 'sport climbers'. He likes to travel through the mountains, climbing as he goes. He was afraid of heights, but overcame his fear, step by step. He climbed out of his comfort zone, as he puts it. 'Did you never simply freeze, going up or down a sheer slope?' I ask. He is often afraid, he says, but overcomes it each time, little by little. He's fed up with Romania and his fellow Romanians, and longs for far-flung climes and climbs.

Above all, he dislikes small-minded, nationalist thinking, he says. He feels like a European, lived in the United States for a while, was in Greece last winter, and swam in the sea in December and January.

The Făgăraş mountains are OK, but he likes Sardinia best, because it's so wild. He's horrified by photos of people queuing to climb Mount Everest.

Home is wherever he happens to be. 'The whole world is my home,' he says. It sounds more like a slogan than the truth. I need something more than that, I tell him. A place to come back to, a centre of the world.

We say goodbye outside the café, and I cross the cobbles to the Black Church – 'black' because of the fire of 1689, which burnt down much of the town, and scorched the walls and collapsed the roof of this ancient building.

At the entrance the quote of the month is from Isaiah 11.6:

The wolf also shall dwell with the lamb, and the leopard shall lie down with the kid; and the calf and the young lion and the fatling together; and a little child shall lead them.

This seems rather appropriate in a Transylvania renowned for its large carnivores and wandering flocks.

The old oak doors are beautifully carved with human figures and spirals. Inside, a woman dressed in black, with long, curly, black hair in a black hood sits, witch-like on a pew in the south transept, her face solemn and staring. Above a side entrance is a mural from the Catholic era, of Saint Barbara handing a rose to Mary – a delicate gesture between young women on a summer's day, though the text explains that the rose symbolises her future suffering.

The church was once the epicentre of the German world in Transylvania, but most of the murmuring voices I hear in it are Romanian. According to the 2011 census, there are only 11,000 Saxons left of a population that once numbered half a million. Close to where the tragic woman sits is the bronze baptismal font, in the form of a chalice, from 1472, thought to be the oldest object to have survived the great fire. All around it on the ground are small, colourfully painted houses made by schoolchildren, with little emblems – *Empathie, Liebe, Hoffnung, Zufriedenheit* – empathy, love, hope, contentment. The pews are carved for the different craft guilds of the city. This church, with its Gothic arches, has been Protestant since it was taken over by the Lutherans in 1543, and the austerity does it good. It's cold inside, though, almost colder than in the streets outside. Along the sides of the church, draped over the back of the pews and dangling from the choir stalls, hang beautiful Ottoman rugs in faded reds, a little out of place.

Braşov was a staging-post on the East–West trade route, and the rugs were once currency here. When Hungary lost Transylvania in 1920, many churches sold their rugs. Saxon churches in villages with few or no Saxons have lent their rugs to this church for safekeeping.

At the foot of the pulpit is a handsome, barefoot Moses. Above him, carved in stone, are the four evangelists, represented by lion, angel, bull and eagle. Above that is a tall tree, with an axe resting on its roots.

Johannes Honterus haunts the church and the public space around it. Born in 1498 in Braşov, the same year Girolamo Savonarola – the Florentine religious reformer who inspired Martin Luther – was executed for contradicting the pope, Honterus became a central figure of the Reformation in eastern Europe. He studied in Germany and Switzerland, where he learnt wood-cutting techniques for printing. Martin Luther published his ninety-five theses in 1517, and was excommunicated by the pope in 1521.

Honterus returned to Braşov in 1539, was elected its first Lutheran pastor in 1544, and established a printing press. His *Rudimenta Cosmographica* or *Description of the World*, published in 1542, included sections on the geography of Europe, astronomy, zoology and medicine. It was published in twenty-six editions, all over Europe, from this southeastern corner of newly Protestant Christendom – such was the thirst of people for news of the wider world, and where they might belong in it. Instead of its usual German name 'Kronstadt', Braşov appears on his maps of Transylvania as 'Corona' – the Latin name indicating 'Crown Town' – which has the same meaning. The crest of the town, visible on all ninety-one of his surviving books, shows the roots of a giant tree that soars upwards, through a crown. Several schools in Braşov bear his name, in honour of his work to make education accessible to the broad public.

The poet Heinrich Zillich attended the Honterus Gymnasium, graduating in 1916. He then enlisted in the Romanian army, and fought the Hungarians for the last two years of the war. After university in Berlin, Zillich returned to Braşov, and in 1923 founded the literary magazine *Klingsor*, which he edited until the outbreak of the Second World War.

Klingsor was a wizard, a magician in German medieval literature, similar to Merlin in the English-speaking world. In the thirteenth-century story *Parzival* by Wolfram von Eschenbach, he is a prince with magical powers, who falls in love with Iblis, queen of Sicily. Caught and castrated by her vengeful husband, he turns his magic against the world of men. By the nineteenth century, Klingsor was rehabilitated, and the German Romantic writers discovered in him a well-spring of poetry and philosophy. In 1919, the novelist Hermann Hesse wrote *Klingsor's Last Summer*, portraying Klingsor as a forty-two-year-old artist, an exile like Hesse, living in Switzerland, contemplating his own death.

In the mid-1920s, Zillich took part with his fellow Saxon writers in a blossoming literary movement with Romanian and Hungarian poets and novelists. The ancient Black Church of Braşov symbolised the Saxon past, and the new railway junction its importance in the present; meanwhile, its theatres and restaurants, backing onto the mountains, established the city as a place where people didn't just pass through, but came on purpose. 'For Zillich the borders of "Europe" reached only as far as the last Gothic cathedral – the Black Church of Braşov,' wrote Réka Jakabházi.

Klingsor published texts from the other languages, as did the Romanian magazine *Abecedar* and the Hungarian *Erdélyi Helikon*.

But politics, especially the rise of Nazi Germany, drove them apart. Hungarian foreign policy worked single-mindedly for Transylvania to be restored to Hungarian control. The poets, laden with their own national baggage, never stood a chance of withstanding that wind, even if they wanted to.

Zillich spoke in the first-person plural of his Saxon identity:

Unser Haus steht frei im Winde der Pässe
Wir sehn die Karpathen himmelumzack

Our house stands free in the wind of the mountain-passes
We see the Carpathians, jagged against the sky.

The Hungarian and Romanian poets wrote of the wind, battering their Transylvanian and national identities. Transylvania seemed different from their own nations, worth preserving for its own sake. Zillich, however, was increasingly drawn to the National Socialist ideology.

Three years before Zillich founded *Klingsor* in Braşov, Sigmund Heinz Landau was born in the same city. It was a time when the Romanian authorities were trying to impose their power. Billboards appeared in public spaces: 'Speak only Romanian', 'You Eat Romanian Bread!' and 'Be grateful! You are a citizen of Great Romania'. As a child, Sigmund remembers a crowd of university students attacking his parents for speaking German, while walking with him in a park. The gendarmes intervened and arrested his parents, instead of their attackers.

He grew up with what he calls an 'impotent rage' against the injustice perpetrated by the peace treaties at the end of the First World War. For him, as for Heinrich Zillich, Adolf Hitler alone offered justice:

By 1932–33 I was a member of the newly-formed and as yet clandestine German Youth Movement, which grew rapidly out of the ranks of the *Wandervogel*, the worldwide German Scout Organisation. By 1935 we appeared on the street, marching and singing in impeccably trained German army style.

In 1940, the first German troops appeared in Transylvania, as allies of Romania. Sigmund found a job as a ground staffer at the airbase at

Ploieşti, north of Bucharest. In April 1941, he took part in the joint German, Hungarian and Romanian invasion of the Soviet Union – in German uniform. He volunteered, and was accepted into, the Waffen-SS. He was home on leave in Braşov in 1942 when an informant with a grudge against him reported him for 'Jewish ancestry'. Thanks to his good record in combat, he was eventually promoted, not executed; but it was close, due to the blind determination among the most zealous Nazis to preserve the 'purity' of their race not just from the Jews, but also from the *Volksdeutsche* – including the Saxons of the Carpathians. Sigmund fought his way through the Second World War, was captured by the Americans, moved to the UK, married an Englishwoman with whom he had two children, and lived to a ripe old age. He wrote a remarkable memoir, *Goodbye Transylvania*.

◎ ◎ ◎

Outside the Black Church, the fog has cleared, but it feels more like November than December. Ciprian picks me up in his car on a street nearby. We're on our way to the southern slopes of the Făgăraş, above Lake Vidraru, where Greenpeace activists are gathering to plan a blockade to stop the logging of the last virgin forests of the southern Carpathians.

The road climbs steeply towards the ski resort of Poiana Braşov, when the sun bursts suddenly through the beeches and spruce that line the road. I feel I'm back in the Black Church, with the sun pouring through the tall Gothic windows, blinding the congregation. That feeling is gradually replaced by a sense of well-being, escaping the gloomy valleys up among the nodding heads of the hills, where the sun always shines. The road is spectacular, winding up between the Piatra Craiului peaks on one side, and Buşteni and Sinaia on the other.

Ciprian was born in 1987 – the year of the Braşov uprising – and grew up with the idea of 'sacrifice'. Inspired by the enthusiasm of those who took to the streets in December 1989 to topple the communist regime, the environmental movement became his adopted tool, to change the world for the better.

After four hours' drive, we reach the Argeș river in Curtea de Argeș, on the southern slopes of the Făgăraș range. In the cathedral here the last king of Romania, Michael I, is buried beside his wife Anna in the family mausoleum.

The Argeș flows thin and stony under a road bridge, robbed of its waters and power by the many dams, storage lakes and hydropower stations upstream. We follow the river up into the mountains, to the Cabana La Cetate guesthouse, just south of Lake Vidraru.

A group of men and women in mountain gear stand chatting outside. The Greenpeace group is distinctive. Other guests come here to relax; Greenpeace is here on a job. One of the activists is Cristian (Cristi) Neagoe, the communications chief. In a few days' time, the latest discussions on the Convention on Biological Diversity (CBD) are due to begin in Montreal, Canada. We talk in the dining room over supper, watched from the walls by the pelts and heads of bears, incongruous among these people trying to save what is left of the forest habitat.

Greenpeace is collecting material to play in Montreal, and is planning a live broadcast from the threatened forests. 'It's one thing to talk about the Amazon rainforest,' says Cristi, 'but it's too easy to ignore the destruction of the virgin forests of Europe.'

The guesthouse restaurant offers Romanian sour vegetable soup, polenta trout (my staple diet in Transylvania) and a rather plain cabbage salad. After dark, one of the Greenpeace teams returns from a recce in the forest. Dorota, a biologist from Poland, has been studying lichens, and has fresh evidence that the forest here has never been disturbed by human intervention. A bit later, Rob announces that it's his birthday, and a bottle of his homemade plum vodka appears, complete with a tray of small glasses. The conversation and the drinks flow deep into the night.

Next morning we set out bright and early in 4x4 vehicles up the mountain. It's still dark and foggy, and I wonder if we will see anything at all.

For the first hour, we follow the Transfăgărașan mountain road. We pass Poenari castle, invisible in the mist, one of many used by Vlad the Impaler in the fifteenth century. Abandoned in the sixteenth, it has been worn down by the cruel Carpathian winters. The new craze for all things Draculean has rescued it from oblivion. According to Rosemary Ellen Guiley's *Encyclopedia of Vampires, Werewolves and Other Monsters*, the vampire hunter Vincent Hillyer spent a night alone in the ruins in 1977, and woke from his nightmares to find puncture wounds in his neck, which had to be treated in the hospital in Curtea de Argeș:

> The doctor seemed divided between whether he should be concerned or amused at my predicament, having been bitten at Dracula's castle.
> 'No, no, no, it wasn't Dracula, it was a spider,' he kept repeating. It must have been a very big spider, because there was about a half an inch between the two wounds. He gave me an antitoxin shot, but I was sick with nausea, fever and malaise for about twenty-four hours. The bite healed in a few days.

The Greenpeace battle with the loggers seems less frightening and more earthly by comparison, with more chance of a happy outcome. We catch a glimpse of Lake Vidraru through the fog as we cross the dam. Then a forest road runs along the Vulcanului creek. The track deteriorates, and we drive gingerly round deep puddles, the mountainside slowly emerging to our left. At one point, a large owl, dark with white patches, flies silently over us from its perch on a fence, up into the forest. The owl is like a watchman, informing the creatures of the woods of our arrival.

Soon there are signs of the loggers: long beech trunks, freshly cut, line the road. The temperature slips to minus two.

We park the vehicles and proceed on foot. It's Sunday, carefully chosen by the activists as a non-working day. They don't want a confrontation until they have a better idea of the forest here. We climb a steep track deeply scarred by loggers' vehicles. One is in place, a red, tractor-like machine with chains behind, to which three beech trunks are attached. Logging began on this plot two months earlier. It's an area first identified on satellite photos by Greenpeace in 2017, and recommended for inclusion in their catalogue of primary forests.

As we climb higher, it starts to look like the site of a massacre. An old maple beside the track is the last tree standing. There's a makeshift fireplace where the workers tried to keep warm during a lunch break. Plastic bottles and food wrappers lie thrown away among the stumps. There's also a stick – a thin piece of timber exactly 3 metres long – which the loggers use to measure the length of the logs they cut, for transporting down the mountain.

The mud is still frozen – so early in the morning, the wintry sun is still too weak to soften it. That makes our progress easier. The sun breaks through the mist below us, through the trees still standing, further down the slope. I look down onto the plains of Wallachia, past the cities of Piteşti and Ploieşti, to Bucharest and beyond, all the way to the Danube, which forms the southern border of Romania.

'One could argue that there is no untouched forest left on earth,' says Ciprian, 'because climate change and pollution impacts even the most remote forest in the Congo or the Brazilian Amazon. So we're not chasing the idea of finding just completely unspoiled forest. We're looking at forests that are valuable and therefore worth protecting.'

The foresters' road ends abruptly in a mess of spruce branches. Beyond that lies the ancient forest.

It's a relief to slip away from the devastation among the folds of the tall trees. We hear birds again, small finches, and ravens cawing a warning of our approach. The first few tall trees have already had a strip of bark shaved away with a single blow of an axe, and a number inked in, starting with 500. Their days are numbered.

The age of the trees varies enormously. Young spruce and beech grow up among middle-aged and grandfather trees. Nothing is crowded. Old trees stand, half broken, or lie spreadeagled on the ground, rotting where they fell. We whisper in awe of the presence of the forest, not wanting to disturb it with our voices.

Valentin, a bearded giant, leads the way with a carved walking stick he uses more for pointing than for physical support.

'This is the Oedipus effect,' he says, pointing to several spruce coming up where the mother tree, now horizontal, once stood. 'They hardly grow for twenty years, in her shade, but when she dies, they shoot up.' The story has a peculiar ring to me, just a week after the passing of my own mother. 'Parents should give their children roots and wings,' she wrote in a memoir. There are plenty of strong roots here, to step between. But it will be a while before I find my wings again.

In one place, a fir has stretched a thick root protectively over the roots and base of a beech. I sit there as the sun rises stronger through the trees, dispelling the mist.

◎ ◎ ◎

The photograph shows a man in his early twenties, Toma Arnăuțoiu, with the caption 'At the King Ferdinand cavalry school in Târgoviște'. By a strange twist of fate, Nicolae and Elena Ceaușescu were executed exactly there, fifty years later. Fresh out of cavalry school, Toma was a handsome officer, hair brushed up from his forehead, large ears, a determined, intelligent look in his eyes, and a square jaw. He looks cut out to be a soldier, while his brother, Petre, has a softer, more oval face, more suited to be a teacher or a poet.

Across eastern Europe, but especially in Poland and Romania, there were sporadic attempts at armed resistance to the communist takeover. Between the elections of 1946 and 1948, anti-communist groups in the mountains tried to make contact with the Romanian National Council in Paris. The communist authorities intercepted the communications and carried out mass arrests in the spring of 1948, nipping any nationwide resistance in the bud. The nationalisation of land, in March 1949, sparked widespread anger, and briefly swelled the number of insurgents. The British consul in Cluj estimated as many as 20,000 anti-communist partisans scattered through the mountains in small groups. But the British and Americans were just observers in this game, not actors or liberators as the partisans dared hope. The coded telegrams home were destined for the dusty shelves of the Foreign Office.

In local parlance, they were referred to as *Haiduci*, rather than partisans, linking them to the folk heroes – the Robin Hood figures – of history and legend. The crackdown by the authorities was brutal, and hundreds were killed in prison or in extra-judicial executions. Thousands of suspects languished for years in Romania's terrible prisons, including Sighet. According to a 1951 report by the Securitate secret police, 804 members of seventeen resistance groups were arrested, of whom less than 10 per cent were affiliated to the former Iron Guard.

In January 1949, Toma was contacted by Colonel Gheorghe Arsenescu to talk about establishing armed resistance to the communists. They assumed that war was inevitable between the West and the Soviet Union. They would establish a base camp in the southern Carpathians, contact other partisan groups in other mountain regions, leaflet the local population, including the police and military, and prepare for the arrival of the Allies. Their concept resembled that of the pro-communist partisans in Slovakia. But while the tide of history was already turning in favour of the Slovak partisans when they took up arms, the Romanians were taking on the might of the powerful young communist state in full flood: no help from outside would ever arrive. Both Toma Arnăuțoiu and Gheorghe Arsenescu had combat experience from the Second World War. Toma had fought the Germans after Romania changed sides in August 1944, while Gheorghe had fought the Soviets, while Romania was still on the German side.

The Arnăuțoiu family came from Nucșoara. In March 1949, twelve men, including Toma's brother Petre, and four women met at the Arnăuțoiu house to make a plan. That night they headed to the hills with light arms – pistols and rifles – and a radio set. Their struggle would last nearly a decade, long after most of the others gave up or were arrested.

The newly formed Securitate was after them from the start. In July 1949, the band split into two groups of six, burrowing crude underground hideouts, distributing typed or handwritten messages to encourage others to resist the authorities, signed 'The National Resistance'. They tore down and defaced the posters of leading communist functionaries outside a lumber station. They also kept notes about their own, hunted lives in the mountains, including close encounters with Securitate troops, and their own poetry.

Finding food, and trying to stay in contact with the Arnăuțoiu brothers' parents and a few trusted friends and priests in the villages, was the most dangerous. The Arnăuțoiu parents were arrested soon after the boys went up into the hills.

Occasionally, the groups would raid foresters' stockpiles. In order not to expose forestry workers to suspicion, they always left a receipt. One listed 60 kilos of flour, 15 kilos of beans and 10 kilos of ham taken from the Berevoiu lumber station.

News of the 1956 anti-Soviet revolution in Hungary offered a brief flicker of hope. There were now only four of the original band left: Toma and Petre, Maria Plop (formerly a maid at the Arnăuțoiu household who had become Toma's wife) and Constantin Jubleanu. That same year, the Arnăuțoiu parents were released from prison and placed under constant surveillance by the Securitate to try to uncover some contact with their sons. An agent 'Mihai' was infiltrated into the village and given a job in the village shop. His task was to befriend and seduce Ana, a girl who worked in the shop who was (rightly) suspected of being in contact with the partisans. To make 'Mihai' seem more credible, the police engineered his detention, along with Ana and 20 litres of brandy, so that they would be locked in a cell together: as a 'partner in crime', Ana might give away vital clues. That line of enquiry withered, however, as it became apparent that Ana didn't know where they were, and she was getting fed up with 'Mihai' anyway.

The secret police suspected several local priests of sympathising with the partisans, and targeted one, Nicolae Andreescu.

The four survivors were now installed in a cave in the 'Fir tree ravine', in a forest a few kilometres from the hamlet of Poenărei. They got down by lowering a rope ladder. They covered the entrance to the cave with a hatch camouflaged with stones and wood, to blend into the rockface. A baby, Ioana, was born to Toma and Maria in 1956.

On the night of 19 May 1958 they were betrayed. The weak link in the chain was Grigore Poenăreanu, an old schoolmate of Toma's who had been identified by the Securitate as a possible target for black-mail, because of his homosexuality and tendency to get drunk and angry. Grigore was occasionally in touch with the partisans, as a source of food and brandy. The Securitate persuaded him to lace their brandy at their next encounter with a powerful drug. They lost their bearings completely, and were unable to resist when troops surrounded the sheepfold where they were hiding. They led the secret police to the hideout in the ravine where Constantin, Maria and the baby were waiting.

The secret police surrounded the area and ordered them all to surrender. Maria lowered the ladder and came down with her baby on her back. Constantin opened fire from the cave and was shot dead by secret police marksmen.

The Securitate archives are copiously illustrated with photographs of all aspects of the resistance: the hideouts, the broken clay pots buried in the ground where they kept their papers, and their pitiful possessions. One picture shows Constantin's lifeless body being lowered from the cave by rope. Another shows Maria, baby Ioana wrapped in a white blanket on her back, her face to the cliff, stepping tentatively down the fifteen steps of the handmade rope ladder. There are also photographs of the weapons cache found in the cave, including four rifles and hand-guns; and of a humble pile of the odds and ends, pans and clothes that sustained their existence for almost a decade.

'We are proud, our heads held high, our foreheads serene. Our consciences are clean because we never robbed. We can never be accused of stealing, though God alone knows how we survived,' wrote Toma.

Toma, Petre and fourteen villagers who had helped them, including three priests, were sentenced to death. The Romanian parliament rejected their appeal for clemency. All were executed on the night of 18 July 1959 in Jilava prison, where Maria and little Ioana were also imprisoned. There are two photographs of mother and child in prison in Pitești, Maria in a dark prison uniform and headscarf, still beautiful, her cheeks hollowed with hunger; her child – now two years old – is staring with curiosity at her captors. Maria died of tuberculosis in prison in Csíkszereda (Miercurea Ciuc) in 1962, as did both of Petre and Toma's parents. Ioana was delivered to an orphanage, and then, aged four, to foster parents; she knew nothing of the true identity of her parents.

After the 1989 revolution, Ioana – then aged thirty-three and a talented violin player and music teacher – became curious to uncover the mystery of her origins. All she knew was the name and address of the orphanage from which she was adopted. 'I rang the doorbell. No one answered. I kept ringing and shouting. Eventually a woman came to the window. I told her who I was. She slammed the window.'

The next day, Ioana tried again. This time, the manager came to the gate to talk to her, but was rather hostile. The woman seemed to know her name. Eventually, she told her where to find the former head of the orphanage.

This time, her reception was very different. The woman burst into tears and kissed her. Piece by piece, Ioana uncovered the whole story. In the early 1990s, the vast Securitate archives were partially opened for public perusal. Ioana discovered a folder with details of the Securitate operations to track down the 'Arnăuțoiu band'.

The last photograph of her father, just before his execution, shows Toma looking older, but defiant. His sister, imprisoned until 1964 and then released during an amnesty, was the only one of her family still alive when Ioana tracked her down.

CHAPTER 21

THE FUTURE OF MANKIND

We must not forget that with these bees we can see into the future
of mankind, that we must live with nature. We cannot cut ourselves
off from it, or twist nature to comply with our demands.

Willi Untsch, beekeeper, Richiş

Willi Untsch lifts the comb delicately from the hive – '*Schau mal...*' – 'Look!' Lemon-yellow. I gaze into the mathematically perfect block of hexagonal cells, oozing with honey, alive with snub-nosed bees. Then he pumps a little more smoke from the burning mushroom that smoulders in his little can. He wears no protective equipment at all – no gloves, no mask; just a broad-brimmed hat to protect his balding head from the September sun. The bees, stoned on mushroom smoke, quieten down. Time stands still. I wonder how it affects their tiny, flowery minds. And my own. The mushroom – tinder fungus, *Fomes fomentarius* – grows mostly on beech trees, and was used in ancient times by those on long journeys, smouldering in their pouches all day long, waiting to light their fires in the evening. The bees are well fed and the weather is still warm, which keeps them calm.

The fine, bright yellow of the tray of honeycombs comes from the Canadian goldenrod flower, *Solidago canadensis*, which contrasts with the white honeycombs, heavy with the pollen of the spring acacia.

The air is rich with the scent of honey. The buzzing of so many bees has a very different effect on the human ear than just a few. It reminds me of the crashing of waves on a shore. Reassured by the mutual respect he seems to enjoy with his bees, I take a step closer. We are in Richiş, a former Saxon village in the Greater Târnava valley, and Willi Untsch is one of the last Saxons, a traditional beekeeper. Romania is one of the biggest honey producers in Europe.

At the end of June and beginning of July there is another lull in the diet of the bees, Willi explains. Most beekeepers feed them sugar to help them survive the winter and the lean weeks of the year; but he trusts them to live off their reserves, gathered earlier in the spring, 'mostly from the dandelion'.

'Look – there is the queen!' I follow the line of his finger. I'm expecting a big bee and am surprised to see a wasp-waisted princess. 'She is smaller than in summer. Her rump is smaller . . . and do you see? She has on her head a louse!'

He holds the queen up for me to look at, and indeed, there is another tiny brown insect clinging to her head. Nursing her between the thumb and forefinger of his left hand, he tries to prize the louse off, between the nail of his thumb and the nail of the forefinger on his right hand. Then he sees a second one on her, poor creature. And he's left his tweezers in the house. No surgeon's hands ever shook less, but he fails. The lice are too small, or cling too tightly to the queen's head. '*Komm her* . . .' he mutters, speaking his deep, heavily accented German to the bugs. 'Come here . . .'

Willi doesn't seem unduly worried. He treats his bees three times a year against the lice, and will do them again soon, in October. The treatment is smoke. The lice fall to the bottom of the hive 'and the bees feel better after that'.

The Varroa mite is another parasite that is devastating the bees of the world, and is a major threat to their survival. The mites weaken the bees by sucking up the fat tissues on their bodies. Propolis, a resinous material made by the bees, is the natural antibiotic of the hive.

Unlike the bees in the spring, who live only five to six weeks, work themselves to death and die, these bees live nine months, through the winter, to May or so, so the population doesn't collapse; some will be born, others die out, and this is a baton which they pass on.

When the new queen makes herself known in the hive with this *kwak-kwak-kwak* noise, the old queen goes to a tree to swarm with the younger bees. With the old queen away, the other young queens do something with their rumps. They give a sound for the other bees – that the next swarm, the after-swarm, should come. A few days later, when you hear this *tu-tu-tu* sound, in contrast to the *kwak-kwak-kwak*, then the after-swarm comes.

I'm not sure I understand everything he says, but I love his explanations, and the sounds he makes, in the language of the bees. I have heard men in the Carpathians utter words in the language of horses, of wolves and birds; but never of bees.

The young queen can fly high, as she is not weighed down with eggs. So she finds a place high in the tree, compared to the first swarm which is lower, because the old queen is full of eggs and has a big heavy rump, so she does not have the stamina of a young queen.

The first swarm, he says, 'is as compact as a rugby ball on the branch, and you can catch it in a bucket; but it can take hours to catch the second one.' Then he sprays it with water, and keeps it for two or three days in a cool cellar. During that time, the bees choose the best of the queens. They sting each other in battle, and the best queens come through.

We walk down the hill, through his garden to the house where he collects the honey. He hands me a big spoon dipped into a whole bucketful of his wildflower honey. The tastes explode in my mouth. It is as if I can taste all 200 varieties of wildflower that grow in King Charles' garden.

Here in Richiş – though he uses the German name, Reichesdorf – nature is still healthy, true and diverse (*gesund, echt und vielfältig*). But when agriculture was still in full bloom here, 'that diversity was of an even greater order, had a different taste to it'.

There were still vineyards, and meadows were cut by hand with the scythe. 'In those days, we had a lot of white clover in August, and wild sage among the vines. But nowadays the sheep have eaten it all. There are too many sheep here nowadays. The small flowers never get a chance to bloom at all.' There are fewer and fewer farmers in the region, he tells me. 'Here in the village, I can count them on one hand.'

Willi sells his honey outside the big Saxon church in Biertan, alongside Ion Cristian's jars. Ion is the Orthodox priest in Copşa Mare and, unlike the static Willi, is an itinerant beekeeper. Each spring, he loads his hives onto a specially designed truck and drives through Romania, following each crop as it flowers. A friend comes with him in his car, so he can shoot back, when necessary, to oversee the Sunday services in his church, if he cannot find another colleague to cover for him.

Ion's travelling season is already over for the year, and I meet him in his garden. Unlike Willi, Ion wears the full beekeeper's white uniform, complete with mask and visor, but no gloves. He lends me his spare set of protective gear, and we wade like astronauts between his hives, freshly landed on Planet Bee. He lifts the combs out to study them, turns them gingerly round to see the honey and the bees on the other side, then slides them back into place like a chess-player.

Ion lights a piece of the mushroom and pumps the bees liberally with its smoke as we walk, clearing a path. His hives are all wooden, painted in pastel shades of blue, yellow and white. Each has a letterbox entry point, around which the bees cluster, funnelling in and out, like liquid insects. The stripes on their backs remind me of tiny tigers.

From Ion's garden I can see the white tower of the Orthodox church where he works; the big Saxon citadel of Copşa Mare at the end of the street; and the reddish tiled roofs of the Saxon houses, spreading through the valley. The roofs seem curved with age, like human spines. The dark greens of summer are still visible everywhere – hardly a hint of autumn, just a pregnant sense and scent of autumn in the air.

Ion is a jovial, priestly man, in a chequered shirt with black waist-coat. The last beekeeper in Copşa Mare. He speaks good English, and there's a rootedness, a peasant determination, about his physical movements and his words. He also has a good range of swear words, even in English, and a wicked sense of humour. Each spring he packs his bees into his small green lorry to drive 500km over the mountains to Calafat on the Danube. He takes the best of his brood. He has a 130 hives, but his truck has room for only sixty-six. 'To make this worthwhile financially, we have to have really strong bee families. If some are missing, you reinforce them with bees from the families that are staying home.' He makes it sound like a sort of Bee World Cup squad.

From Calafat, he follows the acacia trees northwards, as far as Vâlcea, in the bend of the Carpathians, catching the trees as they come into flower. Before he arrives in each village, he rings the local mayor to find out if the farmers have sprayed their crops recently. If they have, he avoids them. In Romania, neonicotinoid-coated seeds are widely used by farmers, though for years they have been banned in the EU. Each year the Romanian government lobbies successfully for an exemption to the ban on such chemicals, which have been blamed for the widespread extinction of bees in many countries. The Agriculture Ministry in Bucharest uses a loophole in the Pesticides Regulations which allows 'emergency authorisations' when a farmer has no other option available to him to save his crop. Romania leads the field in 'exceptions', with far more granted than by any other country. Migrant beekeepers like Ion traverse a minefield.

Another problem is that some farmers say they are using bee-friendly pesticides, but add a little of a more effective but more toxic one when they spray, and that kills the bees. And even if it doesn't kill them, they lose their sense of direction, and the colonies decay.

Ion has a friend who lost 150 hives the previous winter. Badly weakened by sprayed crops, they died in the cold.

'Unfortunately the beekeepers are right – most of the time,' Aurel Enache, a large-scale farmer in Comana, near the Danube told me. 'Sometimes, to get a bigger profit, we overdo the spraying. And we use some products that stay longer in the soil, and are fatal for the bees.' As long as the neonicotinoids are available, he and other farmers will continue using them, he says, because that is what they are used to.

If the situation for beekeepers is so difficult, I ask Ion, why does he carry on doing it? 'Because it's the cleanest form of farming,' he says with a broad grin. 'The only one where you can lick your fingers after work.'

Copșa Mare has a beautiful, disused Saxon church. My friends, James and Rachel de Candole, show me round. They have lived in the former mayor's house in the village since 2015, run a School of Botanical Drawing and ride their horses through the gentle hills of the Greater Târnava valley.

The first time I came here, in the summer of 2019, we stayed in the Ant, a former bee-wagon, converted into a tiny house on wheels. A place to wake early and wade barefoot through a meadow of wildflowers; a place to make a fire by the wagon at night, under a sky unstained by the light pollution of cities.

That July, apprehensive but not scared by tales of the presence of forty bears nearby, we set out for a long walk through the mixed forests of beech and oak. Soon we came on fresh bear tracks. In the damp mud beside them, I made an imprint of my own bare foot. The bear's was almost double in size. The beech trees towered gloriously over us. I wanted to keep quiet in order to see a bear, but was quickly outvoted by my companions, who bellowed and sang so loudly that all the bears of Transylvania must have sought shelter. The tracks continued on our footpath for a while, then turned away, towards a carpet of beechnuts. After a magnificent, but uneventful, hike we encountered two workers from the state gas company, laying a pipeline across an open stretch of

hillside. City men, they advised us to steer clear of the woods alto-
gether: only that morning, they had photographed a small bear crossing
a clearing beside them.

◎ ◎ ◎

The path climbs steeply up above the village to the meadow where
James and Rachel keep their horses. But the horses are nowhere to be
seen.

We find them on the grasslands at the back of the hill. It's a warm,
late-October morning, the temperature close to 20 degrees Celsius, the
leaves of the beech and oak and maple forests just turning. The grasses
are long, except where the local shepherd Manoli's sheep have swept
through. The hills ripple with terraces.

The grasslands and the wildflowers that grow abundantly among
them are what brought James and Rachel here in the first place. As
botanical artists, they are most interested in the flowers; but they are
keen for artists to come to draw and document the grasses, too:

> The challenge is to record not just the flowers in the grasses, but the
> grass in the grassland. These grasslands are essentially fodder for
> grazing animals, the result of hundreds of years of gentle, balanced
> husbandry.
>
> Over the flowering season, the botanist will come out and collect
> specimen plants and press them, label them, and attach them to a
> herbarium sheet. These are stored in our herbarium. Each herbarium
> sheet-dried specimen in our project will be paired with an illustra-
> tion of the same plant drawn from life in pen and wash. With those
> two pieces of information, the dried specimen and the illustration,
> most people will be able to go out into the field and correctly
> identify the flowers, some of which look very similar.

According to Mihaela, a botanist from Sibiu with whom they collaborate, there are some 300 flowering plants within a radius of 1.5km of

the village. At the end, there will be a scientific record of everything growing here – partly for the sheer joy of recording it, and partly as a valuable resource for the future.

The hills we're walking through have changed little in the eight years since James and Rachel arrived. The farming is still balanced and low-intensity, largely because of the hilly topography and the fact that very little of the land has clear deeds of ownership, making it complicated and risky to buy.

In Copşa Mare, an Italian businessman has been snapping up old peasant farmhouses for over a decade, and selling them as holiday cottages without their agricultural land. An old house in poor condition cost perhaps €10,000 five years earlier. It now sells, in the same poor condition, for €30,000. By separating the farmhouses from their farmland, the businessman is cutting off the possibility of a return to the land, were the new owners ever to sell. Each house has a garden, a lawn even, a pretty place to sit for the weekenders. The land in the hills outside the village may one day be sold off to another investor – to enlarge another herd of Black Angus cattle. 'And the whole delicate fabric of this subsistence farming community is shattered,' says Rachel.

Many families in this village have a few pigs and hens, but there are just three or four who farm seriously: notably Stefi, with his thirty water buffalo; and Mihaita, with his six hundred sheep. We can see the buffalo in the distance, black against the yellow-green hillside. When you get closer, and the buffalo notice you when they're lying down, they sort of lean forward on their haunches, in a way cows would never do.

While the shepherd Mihaita walks his sheep through common land pastures, Stefi keeps his buffalo on private land crossed by the new Via Transilvanica long-distance footpath. To stop them wandering away, and keep them safe from bears at night, he has put a simple electric fence round the hilly, grazing area, with signs 'Private property, keep out!', and gates for walkers to pass through. Even so, someone is knocking down the signs, whether because they are indignant that somebody is encroaching on their right to wander, or perhaps in a local feud between farmers, which is common.

James believes those working the land should have priority:

A balance has to be struck. The fences our neighbours put up to protect their buffalo are only temporary. The walkers and folk on bikes should be politely warned that buffalo can be dangerous animals, and to close the gates behind them. Simple!

Small numbers of cattle grazing the land is ideal; but the threat of over-grazing looms large. James cites the case of a Swiss investor in the nearby Hârtibaciu valley, grazing a herd of hundreds of Black Angus cattle, whose meat attracts good subsidies and has become fashionable fare among the Romanian bourgeoisie. On the business's website, it emphasises what it's doing for the environment. In reality, says James, the impact is negative.

The valley stretching some 40km from Agnita to Nocrich is over-grazed, and the amazing diversity of wildflowers and grasses lost. Due to high, robust, electric fencing, a whole swathe of land is placed out of bounds to wild animals. For an investor, fifty head of cattle would be too few: what is needed is hundreds, or ideally thousands. The bigger the better. The investment already affects ten villages. 'This is monoculture. And in such a system, nothing else is possible.'

The key, James and Rachel say, would be to devise policies to keep or restore the balance in the countryside. A sufficient number of small farmers producing healthy food at an affordable price, supplementing their income from farming by renting out a few rooms, and serving good food at their own tables: a state and European Union subsidy system tailored to help local people stay on the land, rather than to support large farmers pollute the land, and land-hungry foreign inves-tors, who funnel away profits to their own shareholders, rather than the local population.

This would also require educating guesthouse and restaurant owners and consumers. They cite the case of a restaurant in the city of Sibiu.

When the owner first opened, her Romanian customers asked for *sarmale* – traditional Romanian cabbage rolls, stuffed with pork and rice. Instead, they started offering locally sourced vegetables, meat, cheese and fish, tastefully presented. And today the Jules restaurant is thriving:

> We don't just need to educate the restaurateurs, we need to educate the customers to ask the right questions, then there's a bigger chance that the restaurants get the message. People should be asking: 'where does this food come from? Was it grown, or raised, or produced locally?'

High up above the village, on Dealul lui Fetea – the Girl's Hill – we study a small, violet flower with yellow pistils and no leaves, on the edge of the meadow. I assume it's an autumn crocus, *Crocus sativus*, but am swiftly corrected: it's a rare Banat crocus, *Crocus banaticus*, with three purple tepals, rather than six.

Near the crocuses there are acacia trees, as tall as my shoulder, spreading out across the grasslands. Wouldn't a few Black Angus cattle grazing here push them back? A few goats would be much more effective, James replies.

> What happens when the investor gets fed up, or gets what he can out of the land, and leaves? In the long term, allowing them to do this is basically writing off these grasslands by saying, 'well, everyone's leaving the land, so we might as well just introduce this monoculture, of single crops, or one kind of animal' – it's very short-sighted. It would not be easy after that to return to a way of farming that venerates the land.

We come to a low fence, put up by a local farmer years before and then forgotten. Even in its heyday, it was easy for deer to jump over, or hikers to climb.

In the next valley, we can see in the distance one of the big apple orchards planted in the communist era, then abandoned. 'The first time we discovered it, on horseback, all the trees were in blossom,' says Rachel. 'It was like entering Sleeping Beauty's garden! It was buzzing with bees and all overgrown.'

In September, cattle and wild animals, including bears, eat the fruit, and locals distil the powerful *ţuică* brandy from what they can scavenge, in an abandoned barn where hops were once stored. We come to a much stronger fence, with large posts driven into the hillside and high-tension cables between them – a sure sign of an outside investor with money. 'At the moment, there are a reasonable number of cattle grazing here, but a year or two ago, there were hundreds. One day, in winter, almost all of them just disappeared, so it's clearly not a local farmer.'

The grasslands are neatly ridged, but it's not quite clear if these are the ridges built by communist pioneers in summer camps, to create terraces for grapes, or simply what the sun, wind and rain have done to the hills over the centuries, like the waves of the receding tide on the sands. James reckons animal hooves and horse-drawn carts have played a role – the horizontal tracks they have made, followed by other animals and occasional humans.

We set out for a five-hour walk along a section of the Via Transilvanica, the long-distance footpath that opened with great fanfare in the summer of 2022, after years of preparation. It starts at the monastery in Putna, up on the Ukrainian border, and winds 1,400km through Transylvania, all the way to Turnu Severin. It is well marked, with capital 'T' signs in a circle etched into andesite stones, or painted on trees and wayside rocks. It has been warmly welcomed by long- and short-distance hikers from all over the world, and James and Rachel are among its fans. They offer long-distance walkers cut-price accommodation in their converted bee-wagon, up on the hillside, where we stayed in 2019.

First we head for Biertan. On the road between the two villages, the bell-ringer from Copşa Mare once encountered a bear climbing over the concrete barrier and up through the undergrowth, past the sandstone cliff. The bell-ringer is paid extra each year to ring the bells of the fortified Lutheran church in Copşa Mare whenever a big storm approaches. On such occasions, he can be seen pedalling wildly on his bicycle to get to the church in time to prevent a calamity of hailstones that would flatten the crops. Is it effective? 'Well, he says it is, and who are we to argue?' replies James. The man points to the storm damage in Biertan as proof of his own prowess – in Biertan there is no equivalent bell-ringer, and therefore no one to protect the crops.

We leave the road, with corn stalks stacked like hay in the fields, to climb up into the beechwood for a beautiful view over Biertan, dominated by its magnificent cathedral. We walk the path children used to take on their way to school.

The rural exodus from the villages, especially of the young, continues almost unabated. On the edge of Copşa Mare we pass the house of a man who organises minibuses to Germany.

Approaching Biertan from the woods, we pause above the Orthodox church. Three elderly women, dressed in black, followed by several younger family members and accompanied by a priest, are weaving their way up through the graveyard, carrying cups and a bottle. They gather round the grave of a recently departed relative. A week or two since the funeral, they've come to say farewell again and drink to a safe journey to the next world. The wine and the ritual helps to spread out and assuage the grief of the bereaved. Their conversation, like the bells from the churches, comes and goes on the breeze. They're discussing how to make the grave prettier, with a small, low wall around it.

Down in Biertan, there are stalls offering local produce opposite the entrance to the church. The wife of Ion, the Copşa Mare priest, is there, selling honey, as is another of James and Rachel's neighbours, selling jam.

In summer, the huge oak tree in the inner garden of the cathedral offers shade. Now the leaves are yellowing. We don't linger – we still have a long walk ahead, over the hills to Richiş.

To rejoin the Via Transilvanica, we walk quite a way out of Biertan in the wrong direction, past the beautifully restored facades of houses, clusters of grapes and dates from the nineteenth century embossed on them. Just one contains a working farmyard, with a parked red tractor and evidence everywhere of farmwork. 'The smartening up of the village brings people in who complain because they don't like cow shit on the road: "we don't want a working farm in the centre of the village anymore. Put it out on the edge."'

It's the same with the smart old people's home, a long, modern building. Until recently, the residents were in an older building in the village centre, where they could go out to the shops, to church, the bar, or just sit on a bench in the sunshine and watch the world go by. But they, too, were moved out of sight of the tourists, like the farms, to the periphery.

We turn left up a track, to cross a meadow.

There's a lone green wagon next to the hedge, with a good collection of buckets and pails outside for the cows, a dog tied to a tree on a chain, and no sign of life. The dog barks at us in that desultory way some dogs bark, knowing its owner will never hear them and the strangers will not be cowed, however fierce the sound. The milking takes place here, beside the wagon. Then the cows are put into an enclosure surrounded by an electric wire fence for the night, to keep the bears away. Each day a truck picks up the fresh milk.

Far up above Biertan, on a wooded ridge, someone has placed a huge Romanian flag, tied between two trees. The track hugs the ridge, the roots of the trees give our feet grip, but also cause us to stumble if we are not careful in the bright autumn sunlight. Beyond the trees, we emerge onto rolling, corduroy grasslands, and beyond them we catch a glimpse of the Făgăraş chain – the jagged teeth sticking up into the clouds, the jaw of the Southern Carpathians.

We are getting deeper into what used to be wine-growing territory, and these are certainly terraces where vines grew. But the vines are long gone, and the long grasses are like a sea, flowing and ebbing over the landscape.

A track bears down from the ridge to the road to Richiş; but faithful to the Via Transilvanica, we bend off to the right instead. The track veers eventually back, through yet another beechwood, till we emerge onto open fields again – a strange sight of reeds and sweetcorn mixed, their yellow stalks almost indistinguishable from each other in the sunlight. There's a slight breeze, and when we stand still, we can hear the reeds and the dry corn stalks whispering and chatting together. The reeds are thinner, and some still fly a banner of seeds like an Ottoman flag, while the maize stalks are fatter. White clouds dot and swell on a dark-blue sky, over the sea of yellow. Even the tall, invasive milkweed, its big pods bursting with dandelion-like seeds, is magnificent. Here you can see how man has shaped these hills and has sown beauty wherever he goes, on purpose or by accident.

Down onto the main road into Richiş, we exchange greetings with three teenage Roma girls who are striking out up a hillside. The middle one is holding her mobile phone up in the air, watching it as she walks. They are not out for an autumnal stroll like us. They've come out like ancient mariners, in search not of nets, but of networks: there is a notoriously poor mobile phone signal in Richiş.

On the approach to the village, the ends of the houses touch the street, then stretch back into what were once vine-clad hills. We knock on a garden gate and Alex appears – the driving force, with his brother, of the Villa Rihuini vineyard. Aged thirty and with two children, he looks impossibly young – perhaps only eighteen. He is fresh-faced, self-confident and an excellent English speaker. There is another family here ahead of us, seated at a long, wooden table in a sort of summer kitchen. We sit down with them. The family is from Bucharest, and the pretty, well-manicured daughter with curly hair and perfect eyelashes introduces herself as an 'influencer'. She has a peculiar laugh. Alex treats us to a surprisingly fruity Riesling, with a pleasant pale-yellow colour and a long, satisfying after-taste. Alex is a great showman, keen to welcome his visitors and promote his wines. He serves fresh bread and *zacuscă* vegetable spread made by his mother, tight-packed bunches of his own grapes, and crisp red apples.

Then he takes me down into his cellar to see the new wines in old oak barrels. Wine-making was once the main occupation in Richiş, and the old houses still have their own cellars. His father bought the farm, and the vines that came with it, from a German family that emigrated in 1990.

After tasting several of his wines, we drive up into the hills to see the vines. We park the car in a field where some people from the city, who 'made their money from prostitution', have built a large, ugly, modern house without planning permission and have tried to block access to Alex's own vineyard. There's no one there, apart from the customary chained dogs. He can hardly wait for the mayor to order them to dismantle it, brick by brick.

He lifts the fence for us to climb under. The vines are looped straight up the poles, a special pruning technique to maximise the yield. All the grapes were picked a month earlier. It's almost November, and the leaves are broad and dying back gloriously, in yellows and shades ranging from green through orange to red. At the end of the rows are autumn peach trees, planted by the diligent Saxons to give an early warning of an attack of powdery mildew on the vines. Where the vines have gone, the peach trees remain. Their long, thin, delicate leaves are turning the same colours as the vine leaves, with an additional shade of blushing pink. Alex and his brother have only 1 hectare. There's only one downside – because of the labour shortage, he couldn't find anyone to clear the weeds between the steep rows, and so he used weedkiller. One result is the hard, compacted earth. Another is the spread of new weeds with black berries – deadly nightshade – the only weed strong enough to resist the poisons.

CHAPTER 22

ON CHICKEN MOUNTAIN

I enter the mountain: a stone gate quietly shuts.
Dream and bridges fly me up.
What violet lakes! What vital time!
The gold fox barks from the ferns.
Holy beasts lick my hands: strange,
bewitched, they stalk with eyes turned inwards.

        Lucian Blaga (1895–1961), 'The Enchanted Mountain'

'Wolf-prints,' says Cristi, crouching beside the path to examine a set of tracks leading off into the trees. Five neat indentations in the snow: four symmetrical toe pads, with claws protruding from them, and a heel.

He explains the difference between the tracks of wolf, wildcat, fox and dog. The cat is the easiest to exclude. All cats, from a domestic kitten to an African lioness, can retract their claws; the wolf cannot, and this is clearly visible in her footprints. That leaves the possibility of confusion with foxes and dogs, which also have similar prints.

One of the beauties of wolves is their sense of purpose: dogs' prints wander all over the place. These prints are very precise and straight, leading down the bank on the far side. They disappear across the main track, blurred by the boot marks of our group, and then pick up again in a sharp line, vanishing among the trees. The toe prints are also longish, while the toe prints of dogs are more rounded. Finally, wolves have bigger feet than foxes. So even though she's a young animal – perhaps nine months old – her prints are already too big to be those of a fox.

It's March and the snow is deep and crisp in the Apuseni mountains. Cristi is leading a small band of intrepid naturalists, including myself, up Chicken mountain for a 'wolf-howling'. Chicken mountain earned its name not as a source of supper for passing carnivores, but because a giant once lived here, who kept a hen who laid golden eggs. The Apuseni mountains include the Ore mountains, famous for their gold mines.

Our plan is to get into position by nightfall, then fall silent to allay the concerns of any wolves disturbed by our approach. Their sense of smell is remarkable: they may smell us, even before they hear us.

The wolves of Transylvania get less publicity than the much-debated bears, but play a no less valuable role in the food chain. The Apuseni mountains are home to perhaps 800 of the 3,000 grey wolves in the Romanian Carpathians, alongside 300 bears. Numbers are difficult to estimate. Different methods are used – wolf-howling (like tonight's planned performance), the methodical tracking of animals, the collection of excrement, and – for the bears – video collars. The plan this winter is to attach three bear collars to individual animals.

The female grey wolf has a gestation period of between sixty-two and seventy-five days. At this time of year, in mid-March, wolves have already mated, so the pregnant females are looking for suitable dens to give birth to their pups. The forest where we are walking is beech and spruce, relatively young, around thirty years old. In this part of Romania, marked by a big exodus from the countryside, there is little road building and the settlements are not expanding. Forestry continues, but there are still large tracts of woodland which are rarely encroached upon. With fewer and fewer humans to disturb them, the wildlife is reoccupying the wilderness.

There are a dozen of us traipsing through the woods, and Cristi is worried that we will frighten away any wolves in the neighbourhood. The roar of mountain streams drowns our heavy footfall, but any self-respecting two-eared beast in the vicinity must have noted our presence.

It's about five in the afternoon and the sparse colours of late winter are intense: the reddish brown of the spruce trunks, the dark green of their needles, the silver of the beech trunks, and everywhere the traces of snow, covering the deep browns of the beech leaves, like litter after a rock festival.

These mountains also shelter a creature that is even rarer than the bear or the wolf: around 200 lynx live here, the shyest animal of all.

The wolves can travel 60km or more in a single night. They hunt in packs of up to a dozen animals. Their prey, most likely a deer, is carefully surrounded, or driven up against a natural obstacle like a mountain ridge or a river. The silent communication among wolves is of almost military precision.

In good spirits, we tramp for about an hour up the track. We try to walk quietly, but the snow is deep under foot, and we stumble and laugh. Some in the group smoke. Then Cristi calls for silence. He divides us into four groups of three, 50 metres or so apart. All whispering, smoking and torchlight must cease. Then we wait and listen.

The mountain and the forest seep back into the space we occupied with all our noise. After what seems an age, Cristi rises silently to his feet, flicks a switch, raises a megaphone above his head, and starts slowly rotating under the stars. A loud, mournful cry – half-man, half-wolf – judders from the throat of the megaphone: first a lone wolf, calling to his pack, then a chorus of others. I detect the voices of three separate animals at least, but there might be more. The howling echoes across the valleys, until I am not sure if the echo is on the recording, from other valleys, or is the real echo of this unearthly chorus. Every wolf, fox, bear, lynx and deer and dog for miles around must be able to hear it.

As abruptly as it started, the howling stops. Cristi lowers himself to a crouch, ape-like, and rests the loudhailer on his knees. We listen, barely daring to breathe, straining our ears in all directions. Nothing. Just the silent roar of the mountain, the stirring of the forest on a moonlit night.

Two minutes later, he rises to his feet and tries again. This time, the sound emitted is more dense and complex, as though of more wolves. And there is something more urgent about it – not so much a male crying out for company, as a conspiracy of wolves, summoning the pack.

We listen. Again, nothing.

Five times, Cristi repeats his curious dance, spinning slowly on a stage in a play about wolves. And five times, there is no reply.

He grins. I see his white teeth. That, too, was a result. Either they are not within earshot, hunting far abroad, or we were too noisy, and the wolves too wise for our tricks. On a normal night – alone or with just one companion – he would then proceed along the tracks he scouted in advance, to a second, then a third site, howling to the wolves until one or two in the morning. But this evening, he will guide us back to the valley, and have a hot drink with us at the guesthouse. He takes the setback in his long stride. Only about 30 per cent of his howling trips elicit a response.

The next day, we drive south, to visit the shepherds and sheepdogs of Brănişca.

Contrary to popular belief, the dogs of Transylvania are more of a menace to man than the bears. Over the years, Romanian shepherds have fallen back on quantity, rather than quality. Romania is a country of stray dogs, and the shepherds' logic is 'the more, the noisier'. The hills and village streets are full of the almost incessant barking of underfed mongrels. There are occasional attempts to castrate or shoot the males. These are met with howls of anguish from dog campaigners and the question is forgotten for a few more years. Entrepreneurs have made Romania one of the top dog-exporting countries in Europe. And

Romanians have got used to rather sad dogs everywhere. In cities, dogs adopt specific blocks of flats, or vice versa.

When you approach a shepherd unexpectedly, up in the mountain meadows, there's a serious risk you might be bitten or mauled by his snarling guard dogs. Some foresters carry a catapult and a collection of small, rounded stones. If a pack of dogs ever go for you, identify the leader of the pack and take aim, one told me. On previous trips, I often carried a stout stick and raised my arm in threat, pretending I was about to hurl a stone. This sometimes works.

◉ ◉ ◉

Brănișca is a village on the Mureș river, northwest of Deva, in the 'ecological corridor' between the Apuseni mountains and the southern Carpathians. The new A1 motorway just to the north of the village was awaited impatiently by drivers for decades, but cuts right across the centre of the corridor. There are reports of bears standing bemused in the middle of the highway, staring down vehicles that are trying to shave thirty minutes off the endless journey to the capital. In July 2023, a driver on the Transfăgărașan highway slowed to let a bear cross the road in front of him. As he filmed out of the window, the bear raised a paw and waved. Millions watched the recording. Where had the bear learnt that?

The *ciobănesc*, or Carpathian sheepdog, is a narrow-faced, intelligent-looking hound, with a grey-and-white shaggy coat. 'When I was first given one,' admits shepherd Ion Cosmin, 'I thought it was useless, because it never barked.' All the dogs he had encountered in his life barked from morning till night. The *ciobănesc* seemed completely docile – more a large lapdog than a fearsome creature that would protect his flock. But he was impressed by how obedient and alert the dog was, and how easy to train.

As they wandered from pasture to pasture, on his first trip up the mountain with his flock and the new dogs, nothing happened for the first few nights. Then one night he heard his dogs bark for the first time. Scenting the approach of a bear or a wolf, the dogs quickly corralled the frightened sheep into one corner of the fenced-off area; then they advanced, snarling ferociously, to confront the threat. 'We call it "selective aggressivity",' explains Radu Mots, a local ranger.

We talk in a meadow surrounded by sheep. Ion and his brother wear big sheepskin coats against the wind that is whipping across the field, and black sheepskin hats.

The conversation is interrupted when a ewe gives birth. We walk over to see. She is carefully licking the blood off the little one, who is already trying to lift her head and stand up. Between tidying up, the mother nonchalantly takes an occasional chew of grass, to fortify herself after the birth.

Breeding pairs of the *ciobănesc* dogs have been given to Ion and other shepherds by Fauna and Flora International and its Romanian partner Zărand since 2018. One condition is that the shepherds only allow them to breed with the other in the pair. Another is that, when puppies are born, these are sold to other shepherds. From such a small start, the aim is to drastically improve the dogs used by Romanian shepherds. Unlike bears and wolves, no one seems to have any idea of how many shepherds the country actually has.

'Many people don't like the conditions attached,' Ion says. He finds it hard to convince local shepherds to buy his puppies. He has had his own dogs for three years, and is very grateful for them. Local and national officials don't do enough to help small, local producers like him sell their goods, he says.

The shepherds are keen to get back to their flocks. We leave them whistling to the sheep, high tones and low, to steer them through their pastures, to the left or the right.

One of the problems for shepherds in Romania since the country joined the European Union in 2007 is that international hygiene regulations can hardly be applied in the high pastures, far from any roads. The shepherds traditionally milk their sheep two or three times a day. The cottage cheese, *orda* and other cheeses from a few days to a few weeks old, take a while to get back down to the valleys.

In 2008, when I visited Poiana Sibiului, the shepherds were up in arms over the regulations, which they saw as an example of urban politicians blindly following the dictates of Brussels, rather than listening to their own farmers. The only loophole seemed to be for shepherds to declare themselves 'traditional small producers', to ease the burden of rules and regulations.

From Brănişca, we drive to Petriş, forty-five minutes to the west. Radu Mots comes with us, to introduce us to a beekeeper, Victor Bulzu. He lives on a hillside with 200 beehives, just above the hamlet of Selište. The track past his house is a favourite migratory route of bears, and bears have a certain fondness for honey.

The year before my visit, one of Victor's neighbours suffered a bear attack on his beehives: eight of the hives were smashed open, the honey gone and the bees lost: Victor and his fellow beekeepers have been given electric fences to keep the bears out.

I climb slowly up the steep hillside beside rows of wooden beehives painted blue, yellow, green and orange, so that the bees can identify their own building. The new fence crackles slightly in the distance.

There's no danger of the bees appearing yet. It's only 6 degrees Celsius during the day, far below the 12 degrees the bees need before they venture out.

Down in the village, Radu shows me the gleaming stainless-steel tanks of honey. I taste each in turn – the linden, the mixed flowers and the acacia, the lightest of them all. They're all delicious, but the mixed flowers is the tastiest – the darkest shade of gold. The pollen from more than a hundred varieties of wildflowers went into this.

There are about forty beekeepers in this region. The Zărand Association initially wanted to help each to bottle and market their honey, but it didn't work. The regulations were too strict, and the cost of all the equipment too high. Now all the producers bring their honey to this small workshop, which streamlines the procedure and makes sure their products get to market – mostly specialised honey and wholefood stores in western Romania.

◎ ◎ ◎

The Apuseni mountains rise from the Transylvanian plateau. Their name comes from the Romanian *apus de soare* – sunset – so whoever dreamt up the name was clearly watching them from the east; perhaps from the less fortunately named town of Turda. It was here in 1568 that one of the great European statements of religious tolerance was issued:

> Ministers should everywhere preach and proclaim [the Gospel] according to their understanding of it . . . no one is permitted to threaten, imprison or banish anyone because of their teaching, because faith is a gift from God.

Three years later, in 1571, a slight amendment was added, just to make sure everyone got the message: 'The word of God shall be preached freely everywhere. No one, neither preacher nor listener, shall come to harm on account of his confession.'

Not far from Brănişca, an ecoduct crosses the motorway from the Apuseni mountains to the Retezat. Like bears, wolves have been known to travel enormous distances. On 1 April 2023, a two-year-old male wolf known as M237 was shot dead by a group of hunters in the village of Hidasnémeti, near the Hungarian border with Slovakia. Born in 2021, one of six cubs in the canton of Grisons in Switzerland, the wolf was fitted with a GPS collar by the Swiss Office of Hunting and Fishing, to find out more about wolf behaviour. At that age, male wolves normally set out in search of a mate from another pack. There were only 180 wolves recorded in Switzerland.

'The dispersal of M237 exemplifies how adaptable wolves are,' wrote KORA, the Swiss wildlife management agency.

> On its migration, it crossed different landscapes, from high moun-tains to cultivated landscapes to settlement areas. He crossed rivers, numerous highways as well as many mountains, one of them nearly

3,500 metres high. Mostly, he wandered purposefully in one direction. Now and then he also stayed for a few days to about two weeks in one place, perhaps to rest, perhaps because of good food supply, before he moved on.

How much further might the wolf have travelled? At what point might he have encountered Carpathian wolves? In Slovakia, the latest estimate of wolf numbers is 300–450. Hunting was banned in 2021.

The person who actually pulled the trigger, according to a Hungarian police investigation, was the nine-year-old son of one of the hunters.

◎ ◎ ◎

The Retezat mountains grow out of the long chain of the Făgăraş range, as the Carpathians curl down to dip at last into the Danube. In the Retezat, they tense their muscles for the last lap of the Carpathian race.

The Retezat was Romania's first national park in 1935, in a wave of enthusiasm for hiking that grew out of the popularity of alpine clubs between the two world wars. There are sixty peaks over 2,000 metres, and just one, Peleaga, over 2,500 metres.

The Retezat 'is quite different from other ranges in the Romanian Carpathians,' wrote James Roberts in *The Mountains of Romania* in 2002, the year of his tragic death. 'A block shaped massif with a landscape of sharp peaks, narrow, boulder-covered crests and tarns filled with natural glacial lakes.'

Driving clockwise round the edge, from Sarmizegetusa in the north, on a summer's day, you get a sense of a great block of mountains, like a pack of cards, shuffling themselves as you glide by, showing one face after another, peering over the ramparts of some Dacian fort.

Ulpia Traiana Sarmizegetusa – not to be confused with Sarmizegetusa Regia, the Dacian capital – is a military stronghold built by the Romans after they defeated the Dacians in AD 107. It's set beside a busy road and boasts a large, smart motel and an open-air restaurant full of descendants of the Romans, tucking into portions so large they may have to hire a room to lie down afterwards.

The base of the walls has been carefully restored, and children run along them and are scolded down by nervous parents. The city was home to 25,000 people at its height, and the strong fortifications suggest that the Romans knew they hadn't killed or enslaved all the Dacians. They must have been afraid of future attacks.

We buy wild raspberries from a young Gypsy woman standing near a temple to the god Silvanus. The price seems high, but it's hot and we don't feel like haggling. But the fruit is delicious. There are compact meadows with small plum trees among the Roman ruins. Silvanus was the god of woods and uncultivated fields.

It's only a short drive now to Brazi, where I've booked a wooden cabin on a hillside above the Râul Mare stream, with a fine view of the cliffs, which rise vertically to the far knuckles of the Carpathians. The owners, a family from Petroșani, fuss around us, making sure we have everything we need. It's a bank holiday and the shops are shut. In fact, we're starving and have not brought any food. They direct us to the restaurant along the road towards the Gura Apelor reservoir. A lot of effort has gone into carving little pockets of civilisation in these mountains – laying on electricity, running water, solar panels and even little pottery figures of angels. But the roar of the stream, the countless stars framed by the mountains, the absence of light pollution – those are the real luxuries.

We set out to scale the Retezat peak, after which the whole range is named. From our perch on the western fringes of the mountains, we choose the entry point at Râușor. Planning ahead for supper, in a small village on the way we buy beans and tomatoes, laid out in trays near the entrance to a garden, at a fraction of the shop price. In Râușor there are several cafés, but it seems strangely impossible to buy a cup of coffee before midday.

We leave the car near the Alpine ski centre and strike out uphill, through a towering forest. The morning is already late, and the forecast is not good – we've somehow chosen the one day when rain is expected. There's also a notable absence of other hikers on this well-trodden path leading up to Lake Bucura, which many use as a base camp. We settle on Lake Ştevia as a more achievable goal for our ancient legs. It's a relief not to be carrying a heavy rucksack with tents, sleeping bags, drinking water and several days' food.

The going is slow, through the great spruce and beech forest of the Ştevia valley. We can hear, but not see, the stream. And somewhere up ahead, we can imagine, but not see, the mountains.

Eventually, we come over a ridge and the trees start thinning out. The path is a narrow trail among large-leaved plants, past tall solitary pines. Then the Lolaia ridge appears directly ahead of us, and the invisible stream from Lake Ştevia emerges to our right. We stop to celebrate under a pine tree on its shore, drinking the ice-cold water straight from the stream. At just that moment, the rain strikes in sudden gusts. We pull on cagoules and shelter in the lee of the rocks.

Round a corner of the path appears a German couple, Siggy and Wolf, soaked to the skin: the rain caught them on the open ground after the forest. They also plan to reach Lake Ştevia, but the weather app on their phones shows a thunderstorm heading our way.

They strike on ahead, determined to reach the lake. We debate the wisdom of continuing, then follow them anyway. The path leads into the middle of a sea of grey-green boulders, and my wife and my sister decide to turn back. The going is getting tougher, the path has disappeared, and red-and-white marks are painted on the boulders. The intrepid hiker can either leap from boulder to boulder, risking a twisted ankle, or feel their way gingerly over.

I press on alone, to catch up with Siggy and Wolf. They're from Munich. She's a teacher, he's semi-retired. Both are seasoned mountain hikers, but are on their first trip in these parts. The wind gets up and the rain is almost constant. We pass the footpath up to the Lolaia ridge on our left. Mount Retezat is almost in front of our noses – a steep, seemingly unclimbable triangle of broken rocks, rising to 2,408 metres. We are just over 2,000 metres now, and at last Lake Ştevia appears to our right, across wet, slippery boulders. The lake is a fierce grey-green, like whipped glass, in a wind which batters my shoulders. I crouch at the water's edge, wash my face in the cold water and cup my hands to drink a little, so I can truthfully say I drank from one of the lakes of the Retezat – from one of the 'eyes of the sea'.

Then we turn and battle our way back, first across the boulders, then eventually into the welcoming arms of the forest. Here, everything is as it was on the way up – no sound of the wind, no stinging sensation of the rain; just the sound of occasional birds and the distant roar of the stream.

More people have appeared, heading our way. Hikers from the central Retezat, from Lake Bucura or beyond, who have fled westwards, towards Râuşor, ahead of the storm. On the way down – only an hour and a half, though my legs are weary from climbing and leaping – we meet one or two intrepid fellows climbing upwards, towards the storm. A bear of a man with a large rucksack and two others, carrying a solar panel and an axe, for the roof of a refuge. Near the bottom, at the foot of the ski slopes, I have a little phone signal for the first time and get through to the girls. They have only just reached the car, after climbing cautiously down through the great forest.

We make it back to our wooden huts at Brazi, with just enough daylight to cook a simple supper, washed down with cans of Ursus beer under the southern Carpathian stars.

◎ ◎ ◎

A huge white truck swings off the main road at Armeniş, onto an unmade-up road. The branches of the trees scrape its sides. We follow in a 4x4, struggling to believe that this lorry – just arrived from Germany with a load of bison – can really make it very far. There is nowhere for a car, let alone an articulated truck, to turn round.

Oana Mondoc is guiding me deep into the Ţarcu mountains, to watch the release of twelve European bison into the wild. Somewhere in the hills ahead, fifty bison are already thriving. Today's arrivals, carefully selected from animals in captivity in parks in Germany and Sweden, are meant to swell that herd. The plan is to reach around a hundred, and then let them be – to trample paths through the undergrowth which smaller animals will be able to use; to eat the bark of shrubs that are overgrowing former pastures; and to reduce the march of trees. And for their own, weaker, members to fall prey to bears and wolves. This is no zoo, not even a protected space.

The steppe bison, *Bison priscus*, once played an important role in tending the great grassland savannahs of Europe, from the British Isles to Spain, and eastwards to the Urals. Between 195,000 and 135,000 years ago, they migrated across the Bering land bridge to become the ancestors of the American buffalo. Larger and more hump-backed than modern-day bison, sketches of them appear in the Spanish cave of Altamira and the Romanian cave at Coliboaia. They are believed to have died out about 5,000 years ago.

The European bison took refuge in the forests of Europe in the Middle Ages, as hunters closed in on them. The rulers of the Polish–Lithuanian Commonwealth issued strict laws to punish poaching. The Białowieża forest in eastern Poland became one of their last places of refuge. During the First World War, German troops shot 600 of them. The last wild European bison in Transylvania was killed in 1790; in Poland in 1921; and in the Caucasus in 1927. That left just forty-eight animals in zoos.

Twelve of those are now the ancestors of all the bison of Europe, nursed back from extinction to reach around 10,000 in Europe today. Three of the herds are in Romania. It's an astonishing and very precise achievement. Now we are bumping along behind them, afraid that at any moment the treasured offspring of Europe's largest land-based mammal might be tipped unceremoniously into a mountain stream after a twenty-hour journey from Germany. Finally, with scratched paintwork, but otherwise intact, the lorry grinds to a halt in a large clearing with space to turn around and reverse. We've come 16km from the main road to Bison Hillock, the release area.

The back door of the truck slams open, a ramp is installed and the photographers gather round. No catwalk has attracted a more eager crowd. But the great beasts are shy. For two hours we wait. The driver gets fidgety. The world's press need to get back to their political intrigues, wars and revolutions. So a ranger climbs up into the truck to find out what is going on.

Curt, a two-year-old male, is to blame. He has placed himself squarely across the exit, blocking the way out for the more sensible females. After much coaxing, he ambles halfway down the ramp, chews on a few leaves, turns up his nose at the paparazzi, and walks back up the ramp to take his place across the exit. Rapid consultations follow among the rangers.

A flap is opened on the side of the truck, from the outside. A gun is brought and poor Curt is shot with a tranquilliser pellet. He crumples to the floor. Eight burly rangers pick him up, two on each hoof, and carry him down the ramp and into a shady corner of the first enclosure, where he is laid gently on a bed of hay. The other bison traipse down the ramp into the enclosure after him.

'On the face of it, we're just releasing some big animals into the forest,' Rob Stoneman of Rewilding Europe tells me. 'But this is a keystone species. They open the forest up.' Rewilding Europe and WWF Romania are in charge of the project to reintroduce bison into the southern Carpathians.

The bison shuffle around peaceably, stretching their legs after the long ride. Rob continues:

This landscape is going to change. What we need is plenty of ecological niches where wildlife can thrive. Because we don't really know how it's going to go: if it's going to be hotter or colder, or more rainy. But it will certainly be more extreme. So we need lots of variety.

The bison are kept in quarantine, in an 'acclimatisation enclosure', for three weeks, then up to eight months in a larger enclosure. Finally, they will be allowed to wander free and make contact with the other animals, released over the previous seven years.

A big decision, Marina Durga, the project manager explains, is not to feed the animals already released, even in winter. This differs from other bison reintroduction programmes. Another decision is not to give them names (Curt was an exception, inherited from his previous keepers). From now on, the bison will be tracked by number, to see what happens to the herd. One or two will be fitted with tracking collars, which send a message every twenty-four hours, so that rangers can keep an eye on their movements. Apart from that, they will be free to wander through a mountainous area of 60,000 hectares. The dream is that one day herds of bison, reintroduced in different areas of the Carpathians, will find each other and mate.

In the village of Feneş near Armeniş, fifty-two-year-old Simona Boieru is stripping the superfluous leaves from her tomatoes. The bison, she says, are 'attracting people from other cultures to visit us, people interested in nature conservation'. According to Oana Mondoc's statistics, 360 people visited Armeniş from abroad in the previous twelve months to see the bison. That's good for Simona, because she supplies tomatoes, beans, sweetcorn, peppers and cucumbers from her garden to the guesthouses where they stay. She also has two cows, three pigs and some hens.

'My first thought when I heard the bison were coming here was curiosity,' she says, showing me which tomato leaves to tear off and which to leave. 'I'd seen them in western films, but never in real life, not even in the zoo. It felt like they were bringing back dinosaurs!'

There were some teething problems at the start, with animals knocking down fences to get into an orchard. But on the whole, the experience has been very positive. 'I have even learnt some English!'

Two streams flow either side of her garden, down from the Ţarcu mountains and eventually into the Timiş river. 'We're used to living with animals. I don't believe animals are more dangerous than humans.' There is a rumble of thunder and then the rain starts, but she hardly notices.

'Our lands here are our homes. If our children go abroad, they'll only live there for a short time then come back.'

Apart from the extra income from tourism, she likes meeting so many strangers. 'These people are really interested in nature conservation, unlike the Romanians!'

She tells me her recipe for nettle tea, which she feeds to her vegetables:

Take a kilo of nettles, and leave it to stew for a week in 10 litres of water. Then take a kilo of the resulting liquid. Mix a litre of this with a litre of unpasteurised cow milk. Add 8 litres of water. Spray it on the tomatoes and on the roots, so it enriches the soil too. You can see from how thick and vigorous the tomatoes are that it's working and protecting them from diseases. There's no need to do anything else.

Two years later, in October 2021, I went back to Armeniş to find out how the bison are faring.

We meet Oana beside a mountain track, at the WeWilder Campus. The wind is howling a near gale, and the building is still under construction. Everything is made of wood – even the traditional shingles on the roof. Inside, everything is pine: the walls, ceiling, floors – like a giant sauna, bathed in daylight from roof windows. Everything in the WeWilder project is carefully conceptualised, to involve local carpenters and farmers.

With Oana we talk as much about the philosophy as the practical details of the reintroduction of the bison. 'For many of us involved here, this is a celebration. People who are preoccupied with the loss and degradation of nature can come here and see how it might be restored.'

Others come for the adventure. To track and witness an animal in the wild, on a sort of European safari.

The main campus building is named 'Zânâ', meaning 'fairy' or 'fairy godmother', a character in Romanian folk stories who rescues children in peril in the forest.

We stay in a tiny house, a wooden cabin on stilts, which belongs to the WeWilder project – a single room with a big glass window, set in an orchard, overlooking the valley. A couple from a nearby farm bring us our vegetarian supper with fresh wild mushrooms, and breakfast the next morning. The cabin's name is 'Muma' – an evil character in the fairytales, the foil to the good fairy, who would boil children alive if they were not rescued by Zână, or by their own ingenuity. We wake up early to the sound of cowbells, and a dew so heavy you could swim in it.

After breakfast, we set out with ranger Matei Miculescu in search of the bison. There are now around a hundred, brought here in groups of a dozen at a time. The females cluster together in herds, already very different from the groups they were brought in. Our plan is to climb up to a ridge above the southernmost area where the bison roam, the Rusca valley. Above us towers Țarcu peak, at 2,192 metres.

We cross a hillside of abandoned apple and plum orchards. There is bear scat beneath the trees in what was once the garden of a mountain cottage. Matei was here tracking a bison herd up beyond the ridge the previous day, but couldn't find them. Bison can walk a long way in twenty-four hours.

We follow one track, then another, up towards the ridge. There's another abandoned shepherd's cottage, its stone walls crumbling into the undergrowth. The building technique was crude, but effective – a base of larger rocks, then smaller stones, fitted in like pieces of a jigsaw puzzle, then a layer of larger rocks, and so on. Clay, which needed to be replaced every few years, was used to stop the northerly wind whistling through the walls. There's a wooden bed in the corner and a beautiful old wooden ladder, all handmade, plus an old horse-hair mattress. The walls are overgrown with rosehips, which the bears also eat, but not until they've finished the apples. The bears need to put on a lot of weight before the onset of winter. They will eat up to 50 kilos in a single day, if they can find it. 'The bears go into these dense orchards and stay there. They feed all day and night. Of course they also rest, but basically they feed non-stop until they finish all the fruit.'

Matei reckons about 70 per cent of the bear population is totally vegetarian, and has never eaten meat. Creatures after my own heart. Those that eat meat are those that have found a fresh animal corpse, and then get a taste for it. Those are the ones that sometimes attack sheep on the high mountain pastures. But more sheep die of disease, or from injuries when they get separated from the herd, or snake bites than from the bears. And the shepherds are used to losing a few sheep, and most of the time don't bother to ask compensation from the state.

The apples are small, reddish and very sweet. We keep going uphill, tramping through dried leaves. We see bear tracks regularly. From the size, Matei reckons it's the same bear. The bear scat is full of pips and apple core. Still a bit moist, maybe two or three days old. 'Looks like the bear ate from the same tree that we did!' says Matei. We also spot canine tracks, either a big dog or a wolf, going the same way.

'There's quite a decent bear population,' Matei says, as we climb. I enjoy the adjective 'decent' here. My sister seems less sure . . . Matei estimates fifteen bears in this valley, roaming all the way up to the top of the mountain. I remember the bears I saw through the telescope in the Maramureş mountains, with the Romanian border police.

WeWilder collaborates closely with the local hunting association, which belongs to Romsilva. 'They've not been shot here since the hunt was banned in 2016. And no one applied for permission to shoot them since then, because we haven't had any bear-human conflicts' – that despite the fact that troublesome bears from other places have been caught and transported here, because of the sheer expanse of wilderness and sparseness of human habitation.

Matei respects local hunters:

The hunters are part of the history, the traditions of this region, passed down from generation to generation. They care more about wildlife than most of the population. They notice the bison, they help us count them, and they really understand their usefulness in the forest.

The bison mating season is September to October, just before our visit. The big males are already roaming alone again, while the younger ones are in groups of two or three. The females give birth every second year, if they're robust and healthy; otherwise every three or four years.

One of the big tests of rewilding the bison was whether the animals would stay in the mountains in winter, or would wander down to the villages in search of food and cause trouble. For the past two years, probably thanks to the mild winters, they have stayed in the mountains – one herd mostly on a pasture above 2,000 metres. In a mild winter they went down to 700 or 800 metres. Earlier incidents in the villages were caused by bison that had not yet adapted to life in the wild and were searching for human company. Those that have adapted know where to find the best food in the hills – grass, coarse vegetation, the bark of trees in the summer; and in autumn, leaves, apples and plums from the abandoned orchards. Shepherds used to pasture sheep and cows here, but fewer and fewer young men are willing to do that. Bears, deer, wild boar and bison have taken over the high mountain valleys.

At last we reach the ridge. Then we set out along a broader track. No more climbing today. There are tell-tale traces – flattened grass where the bison slept, plenty of bison shit, but none completely fresh. Within the herds, the females establish a strict hierarchy. A herd is led by a dominant female, who establishes her position not so much by fighting, as the males do, but by pushing. The herds are also dynamic, one mingling with another. When they separate, a certain exchange of members takes place, as one female finds a better position for herself in the other herd.

We walk for three hours without a single sighting. The bison have become invisible, hidden by Muma (or Zână) from our prying eyes. So we sit on a hillock and open our sandwiches. I've just taken my first mouthful when my sister nudges my shoulder. I follow her eyes. Three big, chocolate-brown bison have just walked out of the forest close by and are ambling straight past us, over the open ground, to another patch of forest. They give no sign at all of acknowledging our presence, and seem completely silent, for such large beasts, like figures from a Native American dream. Then they are gone.

By the autumn of 2023, there were nearly 10,000 European bison on the planet, more than 7,000 of them living free.

CHAPTER 23

RIVERS OF GOLD

They [bracelets] were made of Transylvanian gold, beaten cold then punched and engraved. The utensils used were wooden and metal hammers, covered with leather, wooden anvils, as well as sets of iron chisels and bronze punches.

Romanian National History Museum

In 2016, eleven beautiful golden bracelets were put on display in the Romanian National History Museum in Bucharest. They were strikingly arrayed and subtly lit in royal-purple display cases – spirals of gold with flat, intricately decorated snake's heads at both ends, designed to be worn on the forearm by Dacian princesses (and possibly princes) as a symbol of temporal and spiritual power. From the first century BC, they are proof of the longevity, wealth and skill of the Dacians.

More mundanely, they were the outcome of successful international police cooperation. All had been dug up by treasure hunters in the Orăştie hills close to Sarmizegetusa Regia, the old Dacian capital, in the lawless 1990s and spirited out of the country.

One by one they were found, identified and brought back from collectors and auction houses in Austria, Germany, France, the UK and the US. Under international law, the Romanian state had to pay hundreds of thousands of dollars to compensate the bona fide owners, tricked into buying them by Romanian, German and Serb citizens, who claimed they came from 'their grandparents' collection'.

To prove that they were really of Dacian origin, meticulous laboratory analysis was carried out in Berlin, using X-ray and laser technology. The 24-carat gold, with up to 11 per cent silver and 1 per cent copper, was traced to the gold mines around Roşia Montană and Zlatna in the Apuseni mountains. Gold washed from the Arieş river was mixed with gold mined from seams close to the surface. In July 2022, a Romanian–German citizen, Zeno Pop, got a suspended prison sentence for handling these bracelets and other ancient artefacts. A dozen other accomplices faced fines or prison for their role in the wholesale theft of ancient Transylvanian gold and silver artefacts.

The E79 road towards Roşia Montană winds north from the Transylvania motorway at Deva. The road surface is not good, it's already dark, and occasionally we catch glimpses of mountains growing like mushrooms beneath a canopy of stars. There are hairpin bends, up hill and down dale, and the journey proves longer than we expected. We turn off at Brad; then at Gura Roşiei we go right for the valley of Roşia.

I've reserved a room at a farm which belongs to Eugen David, leader of the Alburnus Maior Association, the environmentalists who, in 2014, stopped a Canadian company, Gabriel Resources, from executing a massive gold-mining project.

Eugen comes out to meet us on the track leading to his home at half past one in the morning, his beard bushier than when I last saw him, at the height of the protest movement. He wears a straw hat – a tall, wild figure waving us into his farmyard. We park beside an old tractor and cart, laden high with hay. The stars are brighter now, with no town or village lights to disturb them.

The next morning, he's up before cockcrow to milk his cows – 'just six now; when you were last here I had ten'. It's mid-August and the long grass in front of his house is full of wildflowers, red clover and field scabious, all purples and greens, soaking my legs in morning dew. Electricity pylons march across his land. There are tamer, no less colourful flowers planted around the farmhouse, a pail of milk by the back door, and any number of cats and dogs to keep his guests company. There's a woman who cycled here all the way from France, returning to the place where she took part in the protests; a Romanian family we managed to wake up with our late arrival; and various other stray guests who have found their way here, off the hard-beaten track. The house looks modern, with creaking wooden doors; but the outbuildings, storing huge stacks of logs seem familiar from my first visit.

The haystacks smell sweet, all the dried flowers of the summer months before us preserved now for the animals this winter. We're above the town, on the hillside of Țarina. Opposite are the mine-scarred hills of Cârnic and Cetate, both sentenced to be beheaded, if Gabriel Resources had had its way. Satellite images show the hills and the Corna valley beyond (where the lake for the tailings – the waste from the ore, once the actual gold is recovered – was supposed to be built) stained white and yellow by two millennia of gold mining. Around the town are artificial lakes. All mining here, before the communists took over, was private, and each family needed access to water to drive their stamping mills. Lumps of ore, brought out from the mines, were crushed to dust for the particles of gold to be washed out. At the height of the gold fever in the late nineteenth century, there were more than a hundred stamping mills.

In the main square, a shop sells sweet buns sprinkled with walnuts. Outside, a woman called Ramona with bright-red hair stands with a huge green parrot called Frida. They fit perfectly in this town and country of extremes. Frida's partner, a male parrot, has flown off in a huff, Ramona explains in a matter-of-fact way. He's perched in a tree near the Catholic church, squawking at her, as she squawks back.

Claudia Apostol, an architect, sits on the outside terrace of the Unitarian church parish house, which she and her team of young architects have restored. Claudia and her husband Virgil were active in the campaign to save Roșia Montană, and stayed on afterwards to help generate work and opportunities for local people. The Greens scared away the investors who 'came to help raise locals from the mud of poverty and ignorance'.

> That's the propaganda, of course. The locals were told 'look, these are the ones that keep your children hungry'. They needed poor people, angry people, to build the mines. But in time some realised that it was all for the sole gain of certain other people they don't even know.

According to the original plans of the Roșia Montană Gold Corporation (RMGC), the gold mined here for the past two millennia had been mined inefficiently. Using state-of-the-art techniques, precious ore could be won from lumps of rock previously cast aside. Furthermore, the houses and churches and graveyards of Roșia Montană (and one or two smaller villages) are built on gold and should be demolished for Europe's largest opencast gold mine to be created. Some 1,800 people were slated for displacement. In the early 2000s, the public relations representatives of RMGC went from door to door, offering to buy the residents out for what sounded like a good price, with the promise of gleaming new accommodation in villages nearby.

From the start there was resistance. In 2005, the five churches of Roșia Montană – Roman Catholic, Greek Catholic, Unitarian, Lutheran and Calvinist – signed a document refusing to sell either their churches or their cemeteries (which RMGC had offered to dig up and move the graves). A hundred house owners refused to sell. But the majority bowed to what they thought was the inevitable, and handed the keys over to the company. They wanted to believe the promises of a better life. Hadn't gold always been mined here? Hadn't powerful chemicals like cyanide always been used? Perhaps the new investments would mean that their children could stay and work in Romania, instead of being forced to travel abroad. Roughly 80 per cent of local people sold out to RMGC.

At the same time, the campaign against the project gathered steam. On 30 January 2000, the brand-new tailings pond of the Aurul gold mine near the northern Romanian city of Baia Mare burst its banks after heavy rains. The pond spilled 100,000 cubic metres of cyanide-laced waste waters, the slurry left over from winning gold, from local mines and from the Roșia Montană region. The poisoned waters flowed into the Lăpuș then into the Someș river, and from there into the Tisa and the Danube. For weeks, fishermen dragged dead fish from the poisoned rivers. Some said the environment would not recover for ten years, others feared a century. But for RMGC, the damage was done. They also planned to use cyanide to separate the 300 tonnes of gold that they estimated lay ready for the taking. The tailings lake they planned would be far bigger than the wounded one at Baia Mare, filling the entire Corna valley. If that broke, opponents pointed out, it would contaminate all the watercourses down to the long-suffering Danube again, and from there the Black Sea.

'Cyanide heap leaching' technology was introduced in gold mining in the 1960s. The cyanide leaches the gold from the crushed rocks, then the gold is separated from the cyanide. The problem is what to do with the residual cyanide and other heavy metals, including the equally toxic arsenic, left over from the process. At the giant Skouries mine in northern Greece, strong local, national and international opposition failed to stop a similar gold mine. At Roșia Montană, the environmentalists won – for the time being, at least.

'The Romanian state tried to modify its own laws, more than twenty of them, to make this private project legal,' Claudia explains. Successive governments – and all the main political parties – were in favour of the project. Most of the locals were won over by the company. But a handful of them – plus determined, well-organised protests, backed by a strong legal team – prevailed because, as she puts it, 'Thank God we are still a democratic state.'

When the [main] law came to a vote, late one night in parliament, there were huge protests in the streets all day, all over the country.

Meanwhile in parliament all the seats were taken, booked two weeks in advance by the mining company.

Protesters in Cluj occupied the headquarters of all the main parties – the Democrats, Socialists and Liberals. And in Bucharest, the ombudsman's office was occupied by protesters. 'There was a short recess between the debate and the vote, when the deputies went out into the hall. And they saw the huge screens broadcasting live from the protests all over Romania.'

Suitably impressed, the deputies simply didn't go back into the chamber. So there was no vote, no law allowing the project to go ahead. But no vote against it, either. It was a very Romanian solution. The project could not go ahead. One by one, all the permits which the RMGC had received, including the permission to destroy the old Roman gold-mining galleries, have been overturned in the courts or have lapsed of their own accord. The jewel in the crown of the resistance was the 2021 decision by UNESCO to declare Roşia Montană a unique landscape, worthy of special protection:

Roşia Montană features the most significant, extensive and technically diverse underground Roman gold mining complex known at the time of inscription. As Alburnus Maior, it was the site of extensive gold-mining during the Roman Empire. Over 166 years starting in 106 CE, the Romans extracted some 500 tonnes of gold from the site developing highly engineered works, different types of galleries totalling 7km and a number of waterwheels in four underground localities chosen for their high-grade ore . . . The site demonstrates a fusion of imported Roman mining technology with locally developed techniques, unknown elsewhere from such an early era.

With a certain poetic justice, the existence of the old mines saved the region from the killer blow of the new.

So, I ask Claudia, why no rejoicing now? Don't you feel that you won? At our feet, Vicki, an ageing French bulldog snores gently. Outside in the yard, young women are sanding down old window frames for churches. Upstairs in the dormitory there is space for thirty architecture students to sleep. Claudia's leg is in plaster from an accident sustained whilst dancing at the summer festival, up on the shore of the Brazi lake, several evenings before.

I would never say we have won. As long as there are 300 tonnes of gold underneath us, we will never win. There will always be pressure here. And it's understandable. This is how the world works, unfortunately.

I hope to live to see the day this community is healed and is aware of what they really have. Their biggest fortune is what they have always had. What their grandparents understood very well. When they won gold from the mountains, they built the churches, the houses, the schools. The gold was used for the needs of this community, not for the needs of some stock exchange investors.

The next morning we visit the Roman galleries. A busload of boisterous teenagers and their long-suffering teachers join us, and we're swept through the gates of the museum, down the steep steps into a cool, dark labyrinth of tunnels. The kids would rather be prizing chunks of gold from the walls, spraying graffiti on them or kissing each other in dark subterranean corners, but we stumble along with them anyway. All too soon, the half-hour excursion is over, and we're back out in the heat of the Transylvanian morning.

Above ground, in a little park, Roman gravestones have been assembled by Claudia and her architects. This is the 'Lapidarium', put together from various Roman-era necropolises in the valley.

'Some of the inscriptions contain not just the names of the deceased, but those of the living, family or slaves, tasked with the upkeep of the graves.' After a gap of 2,000 years, students from Iaşi, Bucharest and Cluj have picked up the work, exactly where the Romans left it.

Tică Darie was another important actor in stopping the mining project. I meet him in the house he has bought and restored, on the main square. Tică is thirty-one and looks even younger – a fresh-faced youth in glasses and with a short beard, glowing with the energy that enabled him to pedal here on his bicycle, twice, from his studies in Copenhagen, to protest against the mine. The FânFest music and theatre festival each summer helped spread the word and the protests.

Like Claudia, Tică was troubled by accusations from local people that the community would be left to rot. A lady in the village gave him a pair of woollen socks she had knitted herself as a present. To raise money for the campaign, he advertised them on Facebook for €5. Within days, he was engulfed with orders for 300 pairs. So he set out through fifteen villages, looking for help. By March, all the socks were delivered, and he had an idea for a business. The coarse local sheep's wool might be good for a one-off emotional appeal, but customers were unlikely to come back for more. What he needed was woollen items of such high quality that he could create a 'Made in Rosia Montana' brand with regular customers. He brought in knitting trainers and designers from Cluj and Zalău, went on a business and marketing course, and even learnt to knit himself, to inspire the ladies.

Large-stitch wool for scarves and jumpers and fine merino wool were imported from Spain and Portugal, spun into yarn in Braşov, and sewn into fine garments in Roşia, first at the ladies' homes and then in his new, well-equipped workshop.

Tică wears one of his own bright-blue merino T-shirts to show us round. First through the wooden gate into his courtyard, in one of the stately homes in the town, once owned by the Bocanicius, a wealthy mining family. His half of the mansion is a buzz of activity. Tică is a scout, and he's set up a branch of the scouts here. Some are camped in his garden, and others are cooking food in his high-ceilinged kitchen. The rubble has been cleared, a terrace built, walls replastered, wooden beams replaced, new windows installed. There's an air of near completion – the white walls glow, the timber smells fresh, and the kitchen is tidy at the heart of the hive.

For a long time, the company thought they could scare me. They threatened the lady who rented me a room, so I offered to buy the whole building. She thought I was joking – I was twenty-three then. But a year later she offered it to me again. I didn't tell her I had no money at all. Within a week, I raised €20,000!

That bubbling enthusiasm has kept him going through all the battles with the company. When they threatened him physically, he sometimes had to scarper – 'running away is not very brave, but sometimes it saves your life'. Later, he found a lawyer in Câmpeni, a nearby town, who helped him stand up for his property-buying plans, and he turned to the local council, even the police, for help against physical threats.

On the day he took possession of the building, company executives and town officials were drinking wine outside the mining exhibition, run by the mining company next door. 'They just watched open-mouthed as I walked into my new home. That was a very symbolic victory.'

By the summer of 2023, Made in Rosia Montana was employing fifty local women, aged between twenty and seventy, and the mockery he faced had turned to grudging respect. Most of the clothes he produces can be delivered anywhere in Europe the next day, and usually for prices below those of the big mountain clothes suppliers. He plans to open a bistro on the ground floor, overlooking the square.

On my last evening in Roşia Montană, I sit down for a beer with Eugen David, my host at the farmhouse and president of Alburnus Major. Eugen has distanced himself from the movement, but I don't understand why.

We sit on the steps of his farmhouse, under a vast umbrella of stars.

'Nothing has changed,' Eugen grumbles. His fellow environmentalists are 'no better than the mining company,' he says, dismissing them as 'neo-goldists', here for their own profit, not that of the community. 'We need the company, which still owns so much of the region, to invest in the local community.'

We climb the Belvedere cliffs overlooking the town. The mining scars on the Cârnic and Cetate massifs are almost hidden. The white towers of the churches catch the morning sunlight. In the far distance, the Vulcan mountain, a flat-topped volcano which lost its head millennia ago, rests like a table, holding up the sky. It's a landscape made by both man and God.

The final stage in the legal battle over the mining project is settled in faraway Washington, DC. In March 2024, the International Centre for Settlement of Investment Disputes (ICSID) ruled in Romania's favour, and threw out the mining company's claim for $4.4 billion in compensation from the Romanian state for blocking the project.

'We will still be here in ten years, in twenty years from now,' Claudia told me. 'Even, in some form or another, one hundred years from now.'

◎ ◎ ◎

In the Petrila district of Petroşani, 150km south of Roşia Montană, Sebastian Tirintica leans on the bridge over the eastern Jiu river. 'It's running clean again,' he says. 'Children can even swim in it. And there are trout, this big . . .' – he gestures with his huge coalminer's hands.

> In my youth, this river was so polluted you could never see the bottom. Here at Petrila they washed the coal, and there was a big pipe pouring out the mining waste, flushed from underground. This was a black river, twenty years ago. So there are some advantages to the coal industry's decline.

Sebastian's father, grandfather and all his relatives worked in the mines: the men below ground, the women above. The Petrila mine was the oldest in the Jiu valley, its lowest shafts a thousand metres deep.

In July 1990, I interviewed miners' leader Miron Cozma here.

Romania's transition to democracy after the December 1989 revolution was rough, and the miners made it rougher still. In May 1990, the National Salvation Front, which shepherded the country through the

– 344 –

first months of the changes, won the first parliamentary election with 66 per cent of the vote, despite pledging that it would disband itself to make way for democracy. Ion Iliescu, the reform communist who once played tennis with Nicolae Ceauşescu, became president. This was more than many Romanians could stomach, impatient as they were to cast off the heavy burden of forty years of communist oppression.

Thousands of students and their supporters set up a protest camp in central Bucharest. The revolution was being stolen before their eyes, they believed.

Faced with daily protests, Iliescu called the miners to Bucharest to 'restore order'. This was the first time I heard the Romanian word *linişte*, meaning 'peace and quiet'. Romania was divided between those who wanted a real reckoning with the communists, and those who, for the sake of *linişte*, thought that the ex-communists would be best placed to oversee the transition. The miners marched through the streets, beating students and smashing up the headquarters of opposition political parties. Seven people died and dozens were injured. This established a tradition of miners coming to Bucharest for political reasons, 'the *Mineriadă*'.

I travelled to Petroşani with two colleagues to interview Cozma. He agreed to talk, on condition that we meet him at the deepest, toughest mine face. So we went down the mine shaft, first by lift, then through tunnels, crawling along narrow passages on our hands and knees. One of the girls fainted and had to be carried to the interview by other miners, delighted that their ruse had worked.

Cozma told us that miners see the world in terms of black and white, because their work is so hard. When the country's president himself said the revolution was in danger, they had no choice but to heed the call, he said.

Subsequent investigations showed that former Securitate agents took part with the miners, and were responsible for some of the worst violence. Cozma was convicted of organising the violence and sentenced to eighteen years in prison for trying to overthrow the established order. He was pardoned by President Iliescu in 2004.

On the bridge over the Jiu river, Sebastian points out the landmarks of an industry sentenced to death.

There were once thirty mines in the valleys, employing 179,000 people at their peak in 1979. The miners were told that enough coal could be found here to keep mining for another 150 years. The coal fuelled communist Romania's massive industrial growth, including its steel industry. There was also a history of protest – the Lupeni mine strike in 1929, when the army opened fire on the strikers and killed at least thirty men; and a protest in 1977 against the Ceaușescu regime, when the government first gave in to their demands, then rounded up the ring-leaders.

After the revolution, state subsidies ended, and many industries which needed coal went bankrupt. Manufacturing and domestic heating switched to gas. The winds of the twenty-first century were against them. In 1990, there were fifteen mines. By 2022, there were just four, at Vulcan, Livezeni, Lonea and Lupeni, employing 4,000 people. There are plans to end all coal mining by 2031.

From the bridge, Sebastian points out blocks of flats six storeys high, built for the miners and their families in the 1980s. They seemed modern at the time, but are now falling apart. In front of several, men are fixing their cars. There are piles of machinery, half overgrown at the roadside.

It's difficult to live here. I worked in Italy for a while, but came home. I was homesick. In retrospect, that was the wrong decision. The minimum wage here is €300 a month [in 2020]. You have to borrow from the bank, there's no possibility to save money for the future of your kids. A kilo of meat in Petrila costs the same as in Germany or Britain, where salaries are seven times higher.

Another problem is that the miners' families were brought from all over Romania by the communists, but in forty or fifty years have still not put down roots. People find it hard to collaborate. The surrounding mountains offer some alternative work in tourism, but nowhere near enough to replace the mining.

In faraway Dobrogea, the easternmost province of Romania on the Black Sea coast, big western wind-turbine companies have begun recruiting men from the Jiu valley, to retrain them as engineers. That is OK for the younger men without families, but Sebastian feels too old to move:

> My older daughter is ten, the younger one five. I cannot imagine they will have a future here. Elsewhere in Europe, there is a painful announcement, and whoever dares to say it pays a political cost. But at the same time, they try to soften the blow, they pay for social programmes, retraining. There's nothing like that here.

For now, the older generation can survive on their pensions. Those who worked for twenty years underground receive between €800 and €900 a month – not bad by Romanian standards.

Sebastian is thirty-four, but only spent five years underground. The mines will not stay open long enough for him to reach that magic twenty-year threshold. One new idea in town is robotics. Perhaps in future they will build gleaming, high-tech robots here, in the shadow of the rusting mineshafts.

We drive up into the hills to get a better view of the valley, 100km long from east to west, eighty from north to south. 'You have to understand,' says Sebastian, 'underground here, everything is coal.'

We look down on the twin towers of Petroşani, the mineshafts and housing blocks of Vulcan, with Lupeni in the distance. Between the tower blocks are the chimneys of thermal stations, which still burn coal to heat the apartments. Weaving between the towns and villages, we make out the eastern and western branches of the Jiu river, uniting to plough southwards, through the mountain gorges, to the Danube. As a child, he says, the snow that fell here only stayed white for a couple of hours, before it turned black.

Every day his wife suggests they go back to Italy. He's not convinced:

I started to feel bad there, mentally. When you wake up in the morning, and you feel no pleasure in going to work, to see anyone.

First I thought we would come back just for a year, but then we stayed. But for me Italy, especially the south, is the most beautiful place on earth.

We follow the eastern Jiu river through the mining villages of Vulcan, Uricani and Lupeni, as ugly as the mountains to the north and south are beautiful. We reach Câmpu lui Neag, a village wiped out in the communist era to get at the coal on which it was built. Fifty-five homes were demolished and the inhabitants moved to the tower blocks of Uricani. When the diggers began, it turned out that the ore was of too low a quality. So they filled the hole with water to form a lake.

On the map, road 66A seems to take us as far as Băile Herculane, our next destination. But as we head west, the road gets narrower and then turns into a rough, impassable track. I ask a Romanian in a four-wheel-drive vehicle coming the other way if I can get through to Herculane, still nearly 80km away. He guffaws good-naturedly and traces a big circle in the air. All the way back to Petroșani, down through the Jiu gorge to Târgu Jiu, then along the southern rim of the Carpathians. Another four hours, with a following wind.

CHAPTER 24

HERCULES AND THE HYDRA

The strangest language is the one of the woods:
The rifle of the night is loaded with star bullets.
The cuckoos will peck the moon to pieces on the tops of alders –
Here grows Antonych – over there grows grass.

Bohdan Ihor Antonych (1909–37), 'Spring'

So, I ask Cristina Apostol, where is the water? She almost shudders. 'It's coming from everywhere!' she whispers. And then I hear it. Flowing underneath our feet. Gushing out of the mountainside. Bursting up among the broken tiles, flowing out from beneath Hercules' helmet, even from the sockets of his eyes. A giggling and gargling of water. And when it rains, it pours down through the walls, off the guttering, drips through the ceiling, and gushes up through the drains.

I'm standing in the magnificent, derelict Neptune Baths, in the ruins of a building where the Habsburg Emperor Franz Josef and his wife Empress Elisabeth, commonly known as Sissy, came to take the waters. The pungent smell of sulphur fills our nostrils. One wing of the building was famous for its sulphurous cures, the other for its saltwater ones.

The first written record of the healing powers of the waters of Băile Herculane, the Baths of Hercules, is from AD 153. Its specular geography, at the toes of the Carpathians, and the porous nature of the limestone on which it stands, has created a miracle of nature. Thermal water gushes out of numerous springs, some as hot as 70 degrees Celsius. People have been drawn here to heal their aching muscles since humans walked upright on the planet. The town became a resort in the mid-nineteenth century. Austrian soldiers injured trying to suppress the revolutions of 1848 came here to heal their wounds.

The Neptune Baths were built between 1883 and 1886, and are now on the brink of total collapse.

From the entrance hall, Cristina leads me down a long, cavernous corridor, painted raspberry red. Everywhere the plaster is peeling, exposing the brickwork beneath. Sunlight from the upper windows splashes on the walls. Over the arch at the end of the corridor are fragments of words in Latin, missing letters: PRO SALUTE . . . PID ET HYC . . . AE.

Creepers push their way through the masonry, their dangling feelers the forerunners of the forest outside, impatient to reclaim the whole building. The floors are covered with dust and chunks of fallen masonry. Shallow puddles reflect ornate ceilings, like an elderly actress gazing in horror at what has become of her body. The building radiates naked-ness, as if the walls had soaked up the bare limbs of all who ever washed here. The erotic power of the baths is overwhelming. Even the young vandals who break in to steal the remaining finery have sprayed huge sprouting penises and vulvas on the walls, and obscene messages.

While the corridor of the sulphurous wing is purple, the salty wing is sky-blue. Cristina leads me into a room with a high broken window facing the mountainside. Brilliant sunlight pours through the top half. A boulder lies in what was once the bath, and more rocks are pushing in through the window frame. Green mould stains the walls. The floor is covered with dry leaves and smashed tiles, and the walls are cracked. Not just the water and the woods are chewing at this building: even the mountains are intent on devouring it. There are thirty-two separate bathing rooms on the ground floor, just big enough for a single bather and their personal attendants.

'We intend to keep one room just like this,' says Cristina, 'to show people the terrible state the baths were once in.' For now, it hardly stands out from the ruins of all the others.

The rectangular marble plaque commemorating where Queen Sissy used to bathe has been stolen, leaving a crude gap. In the bathing room next door, where the emperor splashed about, the plaque is still intact. He was clearly less loved by the populace. Their marriage was long, but not especially happy. Visiting Herculane was one of her great consolations. Sissy was assassinated on the shore of Lake Geneva in Switzerland by an Italian anarchist in 1898 as she was about to board a boat. Luigi Lucheni had a problem with rulers in general. 'I stabbed an empress, not a woman,' he told the judge.

She had been staying at a lakeside hotel, incognito, but a reporter at the local newspaper betrayed her presence. Lucheni, abandoned by his mother at birth and brought up in a succession of brutal orphanages, hoped for the death penalty (which to his chagrin had already been abolished). He hanged himself in his prison cell twelve years later. Sissy was loved and revered by her subjects for ignoring the stiff, arrogant pretensions of Austria's rulers, and for showing an appreciation of the ordinary people and their languages. Some princesses of the twentieth and twenty-first centuries have tried to walk in her bare footsteps.

The Neptune Baths were 'discovered' by Cristina's friend and fellow architect Oana Chirila in 2016. Astonished by their beauty and neglect, she got on the phone. First to the local council, then the Ministry of Culture, then to anyone who would talk to her about them. Her article on the Bored Panda website went viral, and even the prime minister got involved in the discussion. Everyone agrees that something has to be done to save the baths, but by whom, and how?

The fate of the baths since the 1989 revolution is a mirror of Romanian history. In the early 1990s, Valeriu Verbitchi, a wealthy Romanian businessman, bought the building and the old casino opposite for half a million euros, intending to turn it into a playground for Romania's new elite.

In the 2010s, another Romanian businessman, Alexandru Gavrilescu, bought half of it from him. Then the Directorate for Investigating Organised Crime and Terrorism (DIICOT) intervened, alleging corruption.

The building now belongs to the town council, while the land it stands on is the subject of a legal dispute between Verbitchi, Gavrilescu, their children and DIICOT. No one is allowed to restore the collapsing buildings, or do anything to them that would change 'in any way the shape, form, historical essence and structure'.

This is frustrating for the young architects and volunteers. The Hungarian architect Ignác Alpár's original creation was designed to serve the imperial elite of the late nineteenth century. Verbitchi's vision was a continuation of that, at prices the brash new elite of the twenty-first century could afford. By contrast, the Herculane Project proposes a public–private building that would serve both local people and tourists.

The Herculane Project has managed to get the building properly surveyed. Teams have photographed, sampled and assessed every aspect of the walls and roofs, and even the subterranean caverns beneath. They have collected tens of thousands of euros to finance the work themselves.

A door slams in the lofty entrance hall, despite the sultry heat. 'We also have some ghosts,' grins Cristina. A bigger problem is the people who break in regularly.

In the entrance hall, the original fountain is miraculously intact. It is fashioned from majolica, a form of painted and glazed pottery. Cherubic faces curl out of stylised undergrowth; snakes and waterbirds cluster round them, in blues and greens. The fountain was once the only item of colour in the entrance hall.

Outside the main entrance to the baths, a rickety bridge crosses the Cerna. There is a large sign: 'Danger – access forbidden'. The little park opposite was built on the site of a Roman amphitheatre. A giant sequoia tree was planted there in the nineteenth century: now 40 metres tall, it is still rather short for a giant. It feels appropriate to have a sequoia here, to keep the young Transylvanian pines awake at night with tales of the great forests of the Sierra Nevada.

Upstream, there are people bathing in the river, near the house where Sissy used to stay. The young architects have cleared away the jungle of weeds to restore her private walkway, which slopes down to the shore, and have rebuilt the wooden pavilion where she once hung her dressing gowns. They have extended the promenade and built a wooden observatory. In the bed of the river, bathers have arranged circles of rocks, into which hot springs rise and where they can float or paddle. There are five or six, mostly elderly people there today.

'We need to find a way to eliminate the legal obstacles to rescuing the building,' says Cristina, '[including] the written refusal of both individual owners to allow anyone to touch the buildings. That would take an unprecedented intervention by the state.'

We walk through the old square to visit the statue of Hercules. A friendly waitress collects the nationalities of her guests like postage stamps. 'I had a man from Iceland this morning,' she tells me.

On the other side of the street, in the shade of a portico, a man called Josef with a white beard plays Hungarian and Romanian folk dances. There is a pile of 5 lei notes in front of him. After a while, I notice, he tucks those away and places a 10 lei note in his cap instead, in the hope of catching bigger fish.

Hercules rested at Băile Herculane after defeating the nine-headed Lernaean snake, Hydra. This was the second of the ten labours of Hercules. Initially, the poisonous Hydra, who lived in the swamps of Lerna and terrorised the surrounding population, got the better of our hero. Each time he crushed one of its heads with his massive club, two more would spring out in its place. One of the nine heads was, in any case, immortal and therefore could never be destroyed. But Hercules was helped in his labours by his able charioteer Iolaus. Seeing the heads multiply, Hercules asked Iolaus to hold a burning torch to the severed tendons of the neck, each time he struck one off, to prevent new heads replacing it. The ruse worked, until he got to the ninth, immortal head:

> This he buried at the side of the road leading from Lerna to Elaeus, and covered it with a heavy rock. As for the rest of the hapless Hydra, Hercules slit open the corpse and dipped his arrows in the venomous blood.

The geographical Lerna was a city on the east coast of the Peloponnese in Greece, noted for its healing springs and as one of the entrances to the underworld. After his victory, Hercules could have hung around to recover from the battle; but one can hardly blame him for striking out up the Danube, to heal himself here in this shaded valley. His statue in black bronze portrays him bearded and almost naked, muscles rippling, a cloak draped over his left shoulder, his club resting lightly in his right hand. He appears to be strolling towards that friendly café for a healing *pain au chocolat.*

'It was the hour of the post-siesta promenade,' wrote Patrick Leigh Fermor of his own entry to Băile Herculane, on his journey on foot to Constantinople in 1934.

> A band was playing in a frilled bandstand and a slowly strolling throng from Bucharest and Craiova was meandering along the main street, through the gardens, over the Cerna bridge and slowly back again.

The promenaders were dressed to kill, while the British author stank to high heaven.

> Dusty, travel-stained and probably reeking of sheepfolds, I might have been pitch-forked into Babylon, Lampsacus or fifth-century Corinth and as I picked my way through the smart promenaders,

bewilderment was further compounded by an onrush of bumpkin anxiety.

Arriving in the lower town myself, after dark in a hired car, I feel a similar emotion, pitch-forked into Las Vegas, rather than Babylon. The winding road from Tårgu Jiu reaches Herculane through the forests and canyons of the national park, where clusters of parked cars and scantily clad men and women, apparently drunk and in high spirits, spill out onto the road. In place of the pretty, somewhat neglected town I was expecting, we reach a rash of tall hotels, topped by the Hotel Afrodita. Twelve storeys high, like a sore thumb from the Cerna valley, drowning the starlight I hoped for. Whoever granted such a monstrosity planning permission should be condemned to life imprisonment in an under- ground car park. Next door is the equally tall and hideous Hotel Diana, also with 220 rooms and an Aqua Splash machine.

I've booked a room in a small, modern-looking guesthouse. The owner, Florin, runs a gift shop offering unspeakable plastic delights to tourists, but meets me on his bicycle and cycles ahead. We're hungry from the long drive from the Retezat mountains, so I ask Florin for the best pizzeria in town. He directs us to the Crystal, just opposite his souvenir shop. We find a window seat, and an overworked waitress brings us chilled and healing beers. Obese Romanians waddle past, with nagging children fighting over ice creams, water pistols and hydrogen balloons. Elderly grandparents, thin faced and miserable, shout in vain at the kids in their care, to get out of the road before low-slung sports cars mow them down. The pizza dough is crisp, but tastes weird; the topping is generous, but almost completely tasteless. In our hunger, we devour it anyway, then retreat to our room. It's nearly midnight and the disco roar ceases miraculously. We fall asleep, disturbed only by the last of the summer mosquitoes.

In the morning, we get up early to visit the hot-water baths at the foot of the Hotel Roman. It's not quite eight o'clock, but a few regulars are already there. To reach the bathing place, on the shore of the Cerna, you have to tread gingerly through the grounds of the hotel – another astonishingly ugly communist-era creation – then over a rubbish-strewn area to some small, tiled pools. Beyond that, a pipe gushes a constant jet of hot water from the wall, and visitors take turns standing beneath it.

I start in the river, to shock my body with the cold, before treating it to the luxury of the sulphurous heat. The river is swift, but not deep, and as I ease myself down into the waters, I have to grab for support among the slippery underwater rocks. Then I lie back, my shoulders taking the brunt, feet floating free among the rocks. Then I walk to the first rock pool, then the second, the crystal-clear water warmer with every step. There is a jolly camaraderie among the early-morning bathers, a cooing of approval from the old ladies as we take our turns under the natural shower.

There's something incongruous about the industrial-size pipe, the litter on the broken concrete rim, the stubbed-out cigarettes close by, and this magnificent, clean, healing stream and the happy pensioners. I feel my whole body blushing and cleansing in the power of the hot water.

◎ ◎ ◎

From Herculane, we drive down the Cerna valley to the Danube. There's a monumental wall of stone mountains to our left, from Şoimului to Creasta Cocoşului, the Rooster's Crest. Each mountain seems precious now, so close to the end of my journey. I feel differently towards the Cerna now, after testing my strength against hers. Something akin to respect.

To the west, there are mountains of a similar height; then the Cerna starts to get fatter, swollen by the River Danube, which comes up to meet us. The Danube was turned into an obese storage lake by the Iron Gates dam in 1968. Under a scheme agreed between communist Romania and communist Yugoslavia, the river was blocked at one of its narrowest points, to win maximum hydroelectric power. In power-engineering terms it makes sense, and for the past half century has generated significant power for both countries. The price paid by the river, and all the towns and villages on its banks, has been high. The water level in the Danube rose by 30 metres immediately upstream of the dam and continued at an elevated – though steadily decreasing – level all the way to Belgrade, 250km upstream.

The dam blocked the path upstream for the precious sturgeon, which spends most of its life in the Black Sea and only swims up the Danube to spawn.

We pass the turn to Orşova on the left, and keep going. The beautiful old town of Orşova disappeared beneath the waves, and a modern town was built further up the hillside. On a previous visit, the son of the last imam took me out into midstream, to throw flowers into the waters above the lost island of Ada Kaleh. The island was a marvellous remnant of the Ottoman empire, famous for its fig jam, Turkish Delight and pistachio ice cream. A tin of sweetmeats from that era shows a scantily clad maiden with a 1920s hairstyle, languishing among the flowers at the foot of the steps and the motto in Romanian 'Favorita Sultanului' – 'The Sultan's Favourite'.

The road winds along the left bank of the Danube, the Carpathians sloping steeply down to the river over our right shoulders. It feels like the last lap of a very long race. At the start, I sent my mother photographs of each peak and valley, each wildflower meadow. Now it is nearly a year since she passed away.

At Eşelniţa, we park in front of Doru Oniga's Star of the Danube guesthouse. We climb down the stairs to the little restaurant, its wooden terrace jutting out over the river. Then Doru arrives with his son Cătălin, and we order grilled catfish and beer.

After lunch, I climb down the steps and swim out into the green river, like a carp. Doru used to have a huge catfish suspended on ropes beneath the pontoon, but it was washed away by a winter flood.

Soon Cătălin has the speedboat ready, and we're bouncing out over the waters, riding the wake of other boats. In the two hours since we arrived, only one barge has passed, flying a Serbian flag, heading upriver.

First out over the 'little boilers', where jagged rocks in the bed of the river once made navigation treacherous, then to the 'big boilers', while the mountains of Serbia rise tall on the far bank, almost triple the height of the tail end of the Carpathians in Romania. We cross to the Serbian side to look at Trajan's tablet, carved to commemorate his march with a large Roman army to defeat the Dacians in AD 101.

Then back to the Romanian side, to the Mraconia monastery, also built to replace an earlier monastery, now lost below the waterline. There was always a handsome stack of wood beneath the monastery. But that has now been replaced with more accommodation for the nuns. Beside it on the Romanian shore is a long line of navigation signals for passing ships. On the Serbian bank, a large silver ball that could once be raised by a lever system to wave ships through the narrow straits rests in the long grass.

Cătălin guns the boat over to the Grota Veterani cave and we go ashore. I walk barefoot over the hard earth floor, in the half-darkness. There's a hollow in the roof to one side. I sat in the one warm spot in the chill cave, bathed in sunlight here once. This time the light is more diffuse, and each of us explores the cave in silence, lost in our own thoughts. The cave is like the womb of the Carpathians or the tomb at the end of my journey, an ending and a beginning.

I climb back onto the bow and we speed off, just a little further up the Danube, hugging the Carpathian shore. We pass the monument to István Széchenyi, the Hungarian statesman who oversaw the construction of a road in the mid-nineteenth century, and whose engineers blew up the most dangerous rocks, to reduce the danger to shipping.

Our wake laps the sheer limestone cliffs. I scan the crests for eagles. Cătălin cuts the outboard, and we float to a standstill beside the rock. And then we see it. A large water snake, striped and completely motionless, its tail just reaching the water, its head curled around a stone, its head and left eye clearly visible, watching us, unblinking. The Hydra's ninth head, immortal.

FURTHER READING

EPIGRAPH

Milan Rúfus, 'The Quiet Miracle of Motherhood', trans. Allan Stevo. https://www. visegradliterature.net. © Milan Rúfus (heirs) / LITA, 2004/2015.

INTRODUCTION

Albert Wass, *Give Me Back My Mountains*, trans. Susan Tomory, Palmetto Publishing, 2023 (originally published in 1949 as *Adjátok vissza a hegyeimet!*).
Albert Wass: https://en.wikipedia.org/wiki/Albert_Wass.
Endre Ady, 'I am the Son of King Gog of Magog', trans. Adam Makkai. https://mypoeticside.com/poets/endre-ady-poems.
For an excellent guide to long-distance walks through the Carpathians, see Michal Medek, 'Carpathian Thru-hikes'. https://www.transcarpathian.org.
Nick Thorpe, *The Danube: A Journey Upriver from the Black Sea to the Black Forest*, Yale University Press, 2013.
Ernest Gellner, *Nations and Nationalism*, Cornell University Press, 1983.
Anthony D. Smith, *National Identity*, Penguin Books, 1991.
Benedict Anderson, *Imagined Communities: Reflections on the Origin and Spread of Nationalism*, Verso, 1983.
Neal Ascherson, *Stone Voices*, Granta Books, 2003.
Tara Zahra, 'Imagined noncommunities: National indifference as a category of analysis', *Slavic Review*, 69:1 (2010). See also: https://en.wikipedia.org/wiki/National_indifference.
For a discussion of the concept of the Carpathian basin in modern Hungarian thinking, see Péter Balogh, 'The concept of the Carpathian Basin: Its evolution, counternarratives, and geopolitical implications', *Journal of Historical Geography*, 71 (2021). https://www.sciencedirect.com/science/article/pii/S0305748820301237.

The Carpathian Convention was signed in 2003 by all seven Carpathian states: Austria, Slovakia, Hungary, the Czech Republic, Ukraine, Romania and Serbia. http://www.carpathianconvention.org/convention/history.

The Conservation Carpathia Foundation has bought over 27,000 hectares of mountainous forest in the southern Carpathians, in the Piatra Craiului and Făgăraş ranges. A model of forest and wildlife protection: https://www.carpathia.org/who-we-are.

For up-to-date information on the wilderness regions of the Carpathians and beyond, see https://wilderness-society.org.

CHAPTER 1

Johann Herder, *Another Philosophy of History and Selected Political Writings*, Hackett, 2004.

European Commission: https://environment.ec.europa.eu/news/large-scale-study-indicates-wild-bees-are-just-effective-honey-bees-commercial-apple-pollination-2022-12-07_en.

Bálint Varga, *The Monumental Nation: Magyar Nationalism and Symbiotic Politics in Fin-de-siècle Hungary*, Berghahn Books, 2016.

Bryan Cartledge, *The Will to Survive: A History of Hungary*, C. Hurst and Co. Publishers, 2011.

Stanislav J. Kirschbaum, *A History of Slovakia: The Struggle for Survival*, Griffin, 2005.

For the Slovak National Movement: 'The 19th-century Slovak National Movement: Ethos of plebeian resistance', *Nationalities Papers*, 51:6 (2023). https://www.cambridge.org/core/journals/nationalities-papers/article/abs/19thcentury-slovak-national-movement-ethos-of-plebeian-resistance/AEE033B16085639E6B2DC7E1AF8AABA2.

Judit Pál, 'In the grasp of the pan-Slavic octopus: Hungarian nation building in the shadow of pan-Slavism until the 1848 revolution', *Nationalism and Ethnic Politics*, 28 (2022). https://www.tandfonline.com/doi/abs/10.1080/13537113.2021.2004764.

CHAPTER 2

Milan Rúfus: extract from 'The Sleep of the Just', trans. James Sutherland-Smith. https://www.litcentrum.sk/en/sample/poems-milan-rufus. Milan Rúfus (heirs)/LITA, 2025.

Ferdiš Duša: https://craace.com/2019/12/30/artwork-of-the-month-december-2019-hricov-by-ferdis-dusa-1933/.

CHAPTER 3

Tara Zahra, 'Imagined noncommunities: National indifference as a category of analysis', *Slavic Review*, 69:1 (2010).

Susan Gal, 'Polyglot nationalism: Alternative perspectives on language in 19th-century Hungary', *Langage et société*, 136 (2011/12). https://www.cairn.info/revue-langage-et-societe-2011-2-page-31.htm.

Slatinka Association: https://www.slatinka.sk (in Slovak).

CHAPTER 4

For the controversy about Rudolf Vrba's role: https://rudolfvrba.com/vrba-vs-yad-vashem.

Rafting on the Hron river: https://kamnahorehroni.sk/miesta/rafting-hron.

On the Slovak National Uprising: Vilém Prečan, 'The Slovak National Uprising: The most dramatic moment in the nation's history', in M. Teich, D. Kováč and M.D. Brown (eds), *Slovakia in History*, Cambridge University Press, 2011. https://www.cambridge.org/core/books/abs/slovakia-in-history/slovak-national-uprising-the-most-dramatic-moment-in-the-nations-history/BD6945511BA465C3954E41A5854F3C5F.

On the 2004 storm: Mária Havašová, Ján Ferenčík and Rastislav Jakuš, 'Interactions between windthrow, bark beetles and forest management in the Tatra national parks', *Forest Ecology and Management*, 391 (2017). https://www.sciencedirect.com/science/article/abs/pii/S0378112717300361.

Tomáš Hlásny, 'An Orwellian debate on the national parks in Slovakia: What can a scientist do in a post-truth era?', EFI blog, 18 January 2022. https://resilience-blog.com/2022/01/18/an-orwellian-debate-on-the-national-parks-in-slovakia-what-can-a-scientist-do-in-a-post-truth-era/.

CHAPTER 5

For the Oblazy mills: https://www.slovakia.com/sports/hiking/oblazy-mills-in-kvacianska-valley.

For Hanzelka and Zikmund: https://english.radio.cz/when-we-returned-we-were-amazed-how-popular-we-were-legendary-traveller-miroslav-8138642.

On the campaign to save the capercaillie: https://wilderness-society.org/save-the-western-capercaillie-in-slovakia.

On the 'back to the land' movement in communist Czechoslovakia: Edward Snajdr, 'From brigades to blogs: Environmentalism online in Slovakia 20 years after the Velvet Revolution', *Sociologický Časopis / Czech Sociological Review*, 48:3 (2012). https://www.jstor.org/stable/23534999.

We Are Forest: https://www.facebook.com/mysmeles.sk. The forestry industry created a rival platform: https://www.facebook.com/lesysmemy.

For more on the Slovak forests:

https://wilderness-society.org/keep-5-of-slovakia-wilderness.

https://spectator.sme.sk/c/20699965/people-fight-for-vanishing-national-park-forests.html.

https://www.researchgate.net/figure/Two-alternative-forest-reserve-designs-in-Slovakia-a-The-reserve-in-Ticha-and-Koprova_fig3_351065501.

CHAPTER 6

Czesław Miłosz, *New and Collected Poems*, Penguin Classics, 2006.

Kazimierz P. Tetmajer, *Tales of the Tatras*, trans. H.E. Kennedy and Sofia Uminska, Roy, 1943 (originally published as *Tetmajer: Na skalnym Podhalu*, Polish in 5 volumes, 1903–10).

Roman Vishniac, *Children of a Vanished World*, University of California Press, 1999.

Unattainable Earth: Exhibition at the Ethnographic Museum in Kraków: https://etnomuzeum.eu/permanent-exhibition/unattainable-earth.

Jan Brykczyński, 'Boiko: On rural life in the Ukrainian Carpathian mountains'. https://janbrykczynski.com/boiko.

On the Boykos, see also: 'Boykos: the silent inhabitants of the Carpathians', 1 February 2020. https://forgottengalicia.com/boykos-the-silent-inhabitants-of-the-carpathians.

Patrice M. Dabrowski, *The Carpathians: Discovering the Highlands of Poland and Ukraine*, Northern Illinois University Press, 2021.

Stanisław Wyspiański, *The Wedding*, 1901.

Andrzej Wajda, *The Wedding* (film), 1972.

CHAPTER 7

Winston Churchill, *The Unknown War: The Eastern Front 1914–1917*, Thornton Butterworth, 1931. Reproduced with permission of Curtis Brown, London on behalf of The Estate of Winston S. Churchill © The Estate of Winston S. Churchill.

Daniel Mason, *The Winter Soldier*, Little, Brown and Company, 2018.

On the Greek Catholic church at Lutowiska: https://www.lutowiska.pl/church-in-lutowiska.

Timothy C. Dowling, *The Brusilov Offensive*, Indiana University Press, 2008.

Alden Brooks, 'Horrors of the Dukla Pass', *New York Times*, October 1915. https://www.jstor.org/stable/45322639.

Cyprian's *Herbarium* in the Slovak National Museum, Bratislava: https://www.devkid.com/?projects=herbarium.

CHAPTER 8

Johann Herder, *Journal of My Travels in the Year 1769*. https://www.britannica.com/biography/Johann-Gottfried-von-Herder#ref94088.

Paul Magocsi, *With Their Backs to the Mountains: A History of Carpathian Rus' and Carpatho-Rusyns*, Central European University Press, 2015.

For the ancient beech forests of Transcarpathia: https://www.europeanbeechforests.org/world-heritage-beech-forests/ukraine.

On Lesya Ukrainka: https://mypoeticside.com/poets/lesya-ukrainka-poems#google_vignette.

The full text of *The Forest Song*, trans. Percival Cundy: https://tarnawsky.artsci.utoronto.ca/elul/English/Ukrainka/Ukrainka-ForestSong.pdf.

Long-distance hiking trails through the Carpathian mountains in Ukraine: https://www.apacheta.fr/en/guidebooks/great-transcarpathian-trail.

On Alois Zlatnik (1902–79), his life and legacy: V. Kricsfalusy, G. Budnikov and S. Popov, 'Modern state of protected areas in Subcarpathian Rus' (Transcarpathia, Ukraine), suggested by Prof. A. Zlatnik', *Geobiocenological Annals. Mendel University of Agriculture and Forestry (Brno)*, 11 (2007). https://www.researchgate.net/publication/256537059_Modern_state_of_protected_areas_in_Subcarpathian_Rus_Transcarpathia_Ukraine_suggested_by_Prof_A_Zlatnik.

The Deeds of the Hungarians [*Gesta Hungarorum*] is an anonymous twelfth-century manuscript, the earliest written history of Hungary. Hungarian National Library. https://en.wikipedia.org/wiki/Gesta_Hungarorum.

Longo Maï: https://www.thelandmagazine.org.uk/articles/forty-years-longo-maï.

French farmer suicides: https://www.france24.com/en/20181026-suicide-epidemic-plaguing-french-farmers-loire-atlantique.

Svydovets: https://freesvydovets.org/en.

Endre Ady, 'A Hóvár-bércek alatt' ['Beneath the fringe of Hoverla'], first published in the journal *Nyugat* [*West*] in 1909. My translation.

CHAPTER 9

Stanisław Vincenz, *On the High Uplands: Sagas, Songs, Tales and Legends of the Carpathians*, Hutchinson, 1955.

János József Szabó, *The Árpád-Line*, Timp Publishers, 2006.

Nyzhnje Selyshche: https://www.karpaty.info/en/uk/zk/kh/n.selyshche/tastings.

On Béla Franz: https://wilderness-society.org/man-who-cleaned-the-river-environmental-documentary.

Mykhailo Kotsiubynsky, *Shadows of Forgotten Ancestors*, trans. Marko Carynnyk, Ukrainian Academic Press, 1981 (originally published as Тіні забутих предків, Ukrainian). https://diasporiana.org.ua/wp-content/uploads/books/27780/file.pdf.

Sergei Parajanov, *Shadows of Forgotten Ancestors* (film), 1965. https://www.filmcomment.com/article/notes-on-shadows-of-our-forgotten-ancestors-sergei-parajanov.

Martin Buber, *Tales of the Hasidim*, Schocken Books, 1947.

Ménie Muriel Dowie, *A Girl in the Karpathians*, Cassell, 1892.

Maria Sonevytsky, 'Wild music: Ideologies of exoticism in two Ukrainian borderlands', dissertation at the University of Columbia, 2012. She compares studies from the Hutsul region and Crimea. Her book, *Wild Music: Sound and Sovereignty in Ukraine*, Wesleyan University Press, 2019, is a revision of the original dissertation.

Ukraïner periodical: https://www.ukrainer.net/en.

CHAPTER 10

Maksym 'Dali' Kryvtsov, 'My head is rolling from grove to grove', trans. Larissa Babij, Nash Format Publishers, 2024). Maksym posted the poem – Моя голова котиться від посадки до посадки – on Instagram in January 2024, a few days before his death on the battlefront, aged thirty-three. See also https://censor.net/en/resonance/3466231/violets_will_sprout_in_spring_in_memory_of_warrior_and_poet_maksym_kryvtsov. Published with the permission of NF LLC (Nash Format Publisher).

Mykhailo Kolodko: https://kolodkoart.com.

CHAPTER 11

Stanisław Vincenz, *On the High Uplands: Sagas, Songs, Tales and Legends of the Carpathians*, Hutchinson, 1955.

On bears in the Carpathians: http://www.carpathianconvention.org/tl_files/carpathiancon/Downloads/02%20Activities/Large%20carnivores/Conference%20on%20Large%20Carnivores%60%20Protection%20in%20the%20Carpathians/19.10.4.pdf.

Synevyr bear sanctuary: https://synevyr-park.in.ua/en/location/brown-bear-rehabilitation-center.

Julius Strauss:
  https://wildbearlodge.ca.
  grizzlybeardiaries@substack.com.
  backtothefront@substack.com.
Petro Lintur, *A Survey of Ukrainian Folk Tales*, Canadian Institute of Ukrainian Studies Press, 1994.
Illegal border-crossing: see my original report and radio documentary at https://www.bbc.com/news/world-europe-65792384. Some names in this and the next chapter have been changed to protect identities.

CHAPTER 12

Miloslav Nevrlý, *Carpathian Games*, trans. Benjamin Lovett, 2020 (originally published as *Karpatské hry*, Czech, 1982). Available online: http://carpathiangames.org/.
Elie Wiesel, *All Rivers Run to the Sea – Memoirs*, Alfred A. Knopf, 1995.
William Blacker, *Along the Enchanted Way: A Story of Love and Life in Romania*, John Murray, 2010.

CHAPTER 13

Pöang chair: https://www.ikea.com.tr/en/series/poang-series.
IKEA and sustainability: https://www.ikea.com/global/en/our-business/people-planet.
Romsilva is a member of the European State Forest Association (EUSTAFOR): https://eustafor.eu/members/romsilva.

CHAPTER 14

Paul Celan, '[Winter]', in John Felstiner, *Selected Poems and Prose of Paul Celan*, W.W. Norton, 2001.
John Felstiner, *Paul Celan: Poet, Survivor, Jew*, Yale University Press, 1995.
On the Bukovina Germans: https://www.copernico.eu/en/projects/bukovina-germans-inventions-experiences-and-narratives-imagined-community.
Lucian N. Leustean, 'Eastern Orthodoxy and national indifference in Habsburg Bukovina 1774–1873', *Nations and Nationalism*, 24:4 (2018). https://onlinelibrary.wiley.com/doi/pdf/10.1111/nana.12415.
My BBC report with Irina Babich: https://www.bbc.com/news/av/world-europe-60905198.
On the pigments used in the painting of the monasteries: N. Buzgar, A. Buzatu, A.-I. Apopei and V. Cotiugă, 'In situ Raman spectroscopy at the Voroneţ Monastery (16th century, Romania): New results for green and blue pigments', *Vibrational Spectroscopy*, 72 (2014). https://www.sciencedirect.com/science/article/abs/pii/S092420311400068X.

CHAPTER 15

Hutsul horses:
  S. Kusza, K. Priskin, A. Ivankovic, B. Jedrzejewska, T. Podgorski, A. Jávor and S. Mihók, 'Genetic characterization and population bottleneck in the Hucul

horse based on microsatellite and mitochondrial data', *Biological Journal of the Linnean Society*, 109:1 (2013). https://academic.oup.com/biolinnean/article/109/1/54/2415607.
http://heritage-ua-ro.org/en/objects_view.php?id=SV079.
H.S. Timber: https://hs.at/en/index.html.
R. Luick, A. Reif, E. Schneider and M. Grossman, *Virgin Forests at the Heart of Europe: The Importance, Situation and Future of Romania's Virgin Forests*, FreiDok plus, 2021. https://freidok.uni-freiburg.de/data/194387.

CHAPTER 16

Bear mauling: see my BBC report at https://www.bbc.com/news/articles/c8vd26m4r86o.
Bear hide: https://medveles.hu/en.
C. Papp, L. Gál, A. Sallay-Moşoi, I. István and N. Erös, 'Creating bear smart communities: The example of Băile Tuşnad, Romania'. https://cdpnews.net/wp-content/uploads/2024/03/27_4_Papp-et-al.pdf.
Silviu Brucan, *The Wasted Generation: Memoirs of the Romanian Journey from Capitalism to Socialism and Back*, Westview Press, 1993.

CHAPTER 17

Bucura Dumbravă, *Cartea munţilor* [*The Book of the Mountains*], 1920. https://peoplepill.com/i/bucura-dumbrava. A peak was named after her in the Bucegi mountains: Bucura Dumbravă (2,503 metres).
Legends and stories about King Matthias (in Hungarian): https://mek.oszk.hu/06500/06599/06599.htm#58.
Réka Jakabházi, 'Kontaktzonen in der dreisprachigen siebenbürgischen Landschaftslyrik des frühen 20. Jahrhunderts', in I.-K. Patrut, R. Rössler and G.L. Schiewer (eds), *Für ein Europa der Übergänge*, De Gruyter, 2022.
Károly Kós, *Transylvania: An Outline of Its Cultural History*, Szépirodalmi Könyvkiadó, 1989.
Sándor Reményik, *Összes versei*, Auktor, 2000.
Lajos Áprily, *Megnött a Csend – Összegyüjtött versek*, Szépirodalmi Könyvkiadó, 1972.
Lucian Blaga, 'The Enchanted Mountain', in *At the Court of Yearning*, trans. Andrei Codrescu, Ohio State University Press, 1989.
Liviu Rebreanu, *The Forest of the Hanged*, trans. Alice V. Wise, Allen, 1930 (originally published in 1922 as *Pădurea spînzuraţilor*). Rebreanu's brother was arrested as a deserter during the war and executed.

CHAPTER 18

Péter Lengyel: https://peterlengyel.wordpress.com/2012/10/07/hidrocentrale-mici-dezastru-mare.
National Park website: https://www.cheilebicazului-hasmas.ro.
Andrea Tompa, 'Tamási Áron, his plays and the theater'. https://www.academia.edu/7807888/Tamási_Áron_his_plays_and_the_theater.

Ethnic clashes (1990): https://en.wikipedia.org/wiki/Ethnic_clashes_of_Târgu_Mureș.

David Cooper, *Béla Bartók*, Yale University Press, 2018.

The Helikon: https://www.kemenyinfo.hu/en/the-erdelyi-helikon-writers-group/about-the-erdelyi-helikon.

Jaap Scholten, *Comrade Baron: A Journey Through the Vanishing World of the Transylvanian Aristocracy*, Helena History Press, 2021.

Jenő Dsida, *Angyalok citeráján* [*On the Zither of Angels*], Minerva Nyomda, 1938.

George Gömöri, 'Introduction to Jenő Dsida's poems. Poems translated by George Gömöri and Clive Wilmer', *Hungarian Review*, 4:4. https://hungarianreview.com/article/introduction_to_jeno_dsidas_poems_poems_translated_by_george_gomori_and_clive_wilmer.

Tibor Kálnoky: https://transylvaniancastle.com.

CHAPTER 19

Mother Teresa, cited in: Mae Elise Cannon, *Just Spirituality: How Faith Practices Fuel Social Action*, InterVarsity Press, 2013.

Isabel Fonseca, *Bury Me Standing: The Gypsies and Their Journey*, Vintage, 1996.

CHAPTER 20

Petre Arnăuțoiu, permission granted by Ioana Raluca Voicu Arnăuțoiu.

Heinrich Zillich, *Herz der Heimat: deutsche Lyrik aus Siebenbürgen*, A. Langen, G. Müller, 1937.

Ioana Raluca Voicu-Arnăuțoiu, *Luptătorii din munți. Toma Arnăuțoiu. Grupul de la Nucșoara* [*The Fighters in the Mountains: Toma Arnăuțoiu and the Nucșoara Group*], Vremea, 2021. See also https://www.tomaarnautoiu.ro/toma-arnautoiu.php.

Enikő Dácz and Réka Jakabházi (eds), *Literarische Rauminszenierungen in Zentraleuropa: Kronstadt/Brasov/Brassó in der ersten Hälfte des 20. Jahrhunderts*, Verlag Friedrich Pustet, 2020.

Sigmund Heinz Landau, *Goodbye Transylvania*, Stackpole Books, 2015.

Greenpeace: https://greenpeace.at/uploads/2022/11/the-carpathian-forests-report---digital.pdf.

Rosemary Ellen Guiley, *Encyclopedia of Vampires, Werewolves and Other Monsters*, Checkmark Books, 2004.

CHAPTER 21

Lucy Abel Smith, *Travels in Transylvania: The Greater Târnava Valley* (second edn), Blue Guides, 2018.

John Akeroyd, *The Historic Countryside of the Saxon Villages of Southern Transylvania*, Fundația ADEPT, 2006.

Via Transilvanica: https://www.viatransilvanica.com/en.

Copșa Mare: https://www.copsamare.life.

CHAPTER 22

Lucian Blaga, 'The Enchanted Mountain', in *At the Court of Yearning*, trans. Andrei Codrescu, Ohio State University Press, 1989.
Wolves:
> https://www.carpathia.org/study-of-wolves-in-the-fagaras-mountains-using-genetics.
> http://zarand.org/notes-from-the-field-assessing-the-functionality-of-the-corridors-in-the-arad-deva-pilot-area.
> https://www.rufford.org/projects/radu-mot/human-dimension-importance-in-safeguarding-connectivity-between-large-carnivore-populations.
> https://www.kora.ch/en/news/longest-known-dispersal-of-wolf-in-europe--551.
James Roberts, *The Mountains of Romania*, Cicerone Press, 2010.
Janneke Klop, *The Mountains of Romania: Trekking and Walking in the Carpathian Mountains*, Cicerone Press, 2020.
Rewilding: https://rewildingeurope.com/tag/armenis.
WeWild Project: https://www.wewilder.com/team.
European Bison Conservation Center: https://ebcc.wisent.org.

CHAPTER 23

Golden bracelets: https://www.bonadea.net/?page_id=147.
Roșia Montană:
> https://wwf.panda.org/wwf_news/?211153/United-we-save-Rosia-Montana.
> https://www.wmf.org/blog/roșia-montană-birth-movement.
> https://whc.unesco.org/en/list/1552.
> https://www.intellinews.com/romania-wins-6-7bn-litigation-over-rosia-montana-gold-mining-project-316176.
> https://www.researchgate.net/publication/285220489_Rethinking_sustainable_development_of_rural_space_through_the_impacts_of_industrial_activity_A_case_study_of_Petrosani_mining_basin_Hunedoara_County_Romania.
Made in Rosia Montana: https://madeinrosiamontana.com.
Petrila mining: https://www.mdpi.com/2071-1050/12/23/9922.

CHAPTER 24

Bohdan-Ihor Antonych, *The Poetry of Bohdan Ihor Antonych*, trans. Steve Komarnyckyj, Kalyna Language Press, 2016.
Neptune Baths:
> https://herculaneproject.ro/en.
> https://www.europanostra.org/european-heritage-and-financial-experts-visited-endangered-neptune-baths-and-met-stakeholders-in-bucharest.
> https://www.fromrusttoroadtrip.com/blog/baile-herculane.
Lernean Hydra: https://www.perseus.tufts.edu/Herakles/hydra.html.
Patrick Leigh Fermor, *Between the Woods and the Water*, John Murray, 2004.

INDEX